Very Early Clocks

Sari Goldstein Ferber

Very Early Clocks

On Molecular 'Timers' Behind the Origin of Human Experience

Springer

Sari Goldstein Ferber
Department of Psychological and Brain Sciences
University of Delaware
Newark, DE, USA

ISBN 978-3-031-89602-6 ISBN 978-3-031-89603-3 (eBook)
https://doi.org/10.1007/978-3-031-89603-3

This Springer imprint is published by the registered company Springer Nature Switzerland AG
The registered company address is: Gewerbestrasse 11, 6330 Cham, Switzerland

Preface

With treasured memories I thank Heidelise Als and Frank Duffy for guiding my first steps in understanding early behavior. With additional good memories I thank Nava Zisapel for introducing me to the world of chronobiology. With unforgettable memories I thank David Schieber for appreciating my writings and for encouraging me to continue. I also thank Aron Weller and Daniel Yaron for commenting on my steps while I was marching on in the journey of writing this book. Finally, I thank Paul Roos who loved the idea of the book to start with, and Ina Stoeck for being there for me at any time during this endeavor. *If I provided in this book part of the answers, it means that the future will provide the rest.*

Newark, DE, USA Sari Goldstein Ferber

Competing Interests The author has no competing interests to declare that are relevant to the content of this manuscript.

Chapter 1
Introduction: The Intraindividual Perspective – Framing When and How Molecules Become Behavior

Abstract This introductory chapter presents a theoretical framework for this book termed as intraindividual perspective. This theoretical framework relates to molecular and behavioral parallel sequences of embryonic and fetal clockwise development based on novel ultrasonic techniques and recent groundbreaking studies. This theory views the embryo and fetus as a resourceful organism developing from molecular based behavioral processes towards the emergence of psychological phenomena around mid-gestation and from then onwards. This gestational progress unravels time-specific underlying molecular combinations, which accounts for intra- and extrauterine environments too. This theory, aimed at outlining the pathway between the genome and the changes in early behavior, argues that molecular events elicit behaviors at specific times and develop in a combinatorial manner through a series of zero to one binary step responses. This theory also argues that in healthy gestation the molecular building blocks of fetal behavioral achievements develop in parallel sequences of combinatorial series of binary processes on the vector of time. These parallel sequences develop from large levels of entropy implying simplicity in behavioral outcomes in the embryo through reduced entropy and increased free energy from mid-gestation onwards, with growing complexity in behavioral outcomes and their molecular building blocks late in gestation. Finally, this theory argues that reduced entropy facilitates reoccurrence of a behavioral achievement once it occurred through reversible neurobehavioral states in the prenatal thermodynamic system of the human organism, historically termed by Boltzmann as reversible macro-states. This theory embraces all other chapters of this book, raging from clocks of genetics, epigenetics, myelination, and antioxidative defenses as inputs, to clocks of self-regulation, motivation, emotions, reward experiences, perception, and executive functions as outputs. The inputs and outputs represent the complex of very early clocks working in a 40-week changing thermodynamic system via timewise alterations in combinations rather than through a simple linear progress. The concluding chapter presents the concluding remarks and modeling of this theory by showing how clock genes are designed as "timers" to dictate optimal ranges of time for other very early clocks. It is concluded that clock genes design the time of available free energy for any onset of behavioral accomplishment by inducing molecular sensitivity to intra- and extrauterine stimulation

S. Goldstein Ferber, *Very Early Clocks*,
https://doi.org/10.1007/978-3-031-89603-3_1

after mid-gestation, beyond their photic reactivity and for shorter and longer timeframes than their known 24-hour cyclicity.

How many times have you tried as a scientist to follow and unravel the complex neurobiological underlying mechanisms, which you suspected are responsible for the psychological birth of the human infant? How many colleagues argued with you that psychological phenomena may be detected before birth? How many times have you wondered how the first step, the first word, the first wish of your child has occurred and whether it happened on expected timelines? How many times have you wondered if your child just knew how to do it and not necessarily in a manner related to your parenthood? Indeed, scientists are fascinated for decades by the wonders of early development. Equally, parents and families derive joy and happiness on every first accomplishment of their children. Where does it all begin?

This book will suggest that psychological phenomena are part of embryonic, fetal, and infant development and that they are activated into behavioral and mental experiences by biological clocks. This book presents the molecular building blocks as clockwise inputs and the psychological early behavioral accomplishments as early outputs. This book will relate these biological clocks to the early functioning of clock genes. Finally, these very early clocks will be presented as input-output complexes within models of control systems adapted from the fields of neural networks, mathematics, and engineering. Part I of the book is dedicated to the inputs and Part II—to the outputs. Unlike the data shown in Part I, the molecular data in Part II is specific to the psychological phenomenon in each chapter. The integrating concluding chapter presents a control systems model using logic gates and presents clock genes as the master controller of time-dependent molecular parallel sequences which govern the rise of psychological phenomena during gestation.

This introductory chapter will present the intraindividual perspective behind this book. The intraindividual perspective on development presents the fetus and infant as a preprogrammed, clockwise, capable, and organized capsule, including its sensitivity to environmental stimuli. This growing capsule is presented on molecular levels at predefined times along gestation, showing a profound peak at mid-gestation.

The unique view of the intraindividual approach lies in the focus on the fetus rather than on a dyad (e.g., pregnant mother and her fetus, parent and infant), viewing him/her as a unit that manages internal and external cues. This is a different focus on the growing individual, relating individuality to earlier stages of development than articulated to date. The link between molecular events and behavior on the vector of gestational times is the challenge presented in this book, addressed as very early clocks.

Thus, according to the intraindividual perspective, individual strengths and thresholds for mental experience and behavioral functioning are generated very early in embryonic life by molecular events, and they continue to develop through fetal, early infancy, and later life. This conceptualization of early human capacities and behavior suggests that underlying molecular processes determine a range of potential functions on an axis between maximum potential strengths and maximum potential thresholds to reach an accomplishment of a developmental milestone. Maximum potential thresholds attenuate the maximum potential strengths. This theory argues that reduced molecular entropy facilitates reoccurrence of a behavioral achievement once it occurred through reversible macro-states in the prenatal thermodynamic system of the human organism, historically known as Boltzmann's view. This approach, first appearing in several of my recent publications (Ferber et al., 2022, 2023a,b), represents a comprehensive understanding of the prenatal psychological phenomena recorded in the last decade by 3D and 4D ultrasound techniques and shown in groundbreaking prenatal studies of the fetus.

To outline the general time-sensitive pathway responsible for the rise of behavioral complexes, it is noted that a feedforward order is apparent in the consolidation of, first, the autonomic system upon the closure of the neural tube at 28–32 days of gestation, then the motor system upon the onset of sensorimotor capacities at 11–13 weeks of gestation, and then, the attention system upon the emergence of sleep state discrimination, maturation of the visuomotor system, and the onset brief alert states at 27–31 weeks of gestation. The interplay between the systems is being consolidated into the availability of the early mental capacity in healthy gestational progress by coupling of movement-sleep states and heart rate around mid-gestation (Donovan et al., 2020; Einspieler et al., 2021; Fagard et al., 2018). More specifically, the level of oxygenation determines the maturation of ANS synchrony (Ferber et al., 2022). ANS synchrony triggers the maturation of sensory and motor fibers. Maturational connectivity between sensorimotor developing regions triggers the resolution of the subcortical plate, the appearance of the first cortical functions, and its control over the brainstem as very early attention capacities (Ferber et al., 2022, 2023b). The feedforward process is dependent on the timely interplay of an earlier developing system with the newly developing one. Indeed, this may seem a simple linear progress of development. However, I show in this book that this feedforward order is developing through many fluctuations along development, then a peaking phase, and only thereafter a new order of interplay between the new and the old systems, better described as a step response.

That is, any new behavioral accomplishment of the fetus occurs on the basis of the latest system interplay in a zero to one manner followed by a plateau over the time that elapses until the next accomplishment and "jump" for establishing a new interplay between the new and the older systems, while incorporating the new system into the behavioral repertoire. The systems work together, influencing each other (Ferber & Makhoul, 2004), towards a coordinated balance by increasing or decreasing responses to each other's state, starting from early gestational age. This creates a combinatorial curve, rather than the traditional simple linearity (Arlotta et al., 2005; Kolk & Rakic, 2021). I show in this book that the binary molecular combinations of progenitors' expression vs. non-expression, their resulting cell differentiation vs. non-differentiation, and cells' fixed migration following target location vs. non-migrating cells, reaching binary neural polarization vs. non-polarization, are processes that do occur in a time-specific, combination-specific manner, rather than in an area-specific simple localization or function. Thus, fetal behavioral development, and its molecular basis too, can be understood only as combinatorial phenomena of binary processes, such as step responses on the vector of time.

The available span, for each molecular event and its behavioral output, meaning each "jump" or step response, is termed here as dependent on the neural free energy vs. levels of early cerebral entropy. The highest levels of entropy in the brain have been found in early life (Scahill et al., 2024) and aging (Cieri et al., 2021; Tatullo, 2024) and were therefore historically related to reduced complexity and increased simplicity. Recent neurological studies found the inverse relationship between cerebral entropy and signal complexity in very-early-born preterm infants representing validation of the historical view (Ferber et al., 2024). In my terms, the onset of optimal levels of free energy versus entropy, result in more complex stimulus-related fetal behaviors and the onset of short transitory alert states, which develop during embryonic life and clearly emerge at mid-gestation. This emergence appears following an excitation period from conception to 20–24-week gestational age through the interchange of excitation role between the early excitatory GABA and its replacement by the excitatory of glutamate at mid-gestation, including the molecular cascade resulting from this interchange (Ferber et al., 2022; Herlenius & Lagercrantz, 2004).

To unfold this complexity of the excitation vs. inhibition and entropy vs. levels of free energy, the Boltzmann's theorem is suggested here to define systems in thermodynamic equilibrium as having a distinct probability of creating a corresponding microstate of available energy. The sum of microscopic states is available to the system dependent on macroscopic limitations, meaning the system's level of entropy (Ferber et al., 2024). It has been suggested that once a system has reached its thermodynamic equilibrium, it has maximized its entropy and no further change is possible (Goldstein & Lebowitz, 2004). Additionally, it has been recently suggested, as an understanding of the second law of thermodynamics regarding increased disorganization over time and in relation to the Darwinist order of selection of species (Darwin, 1859), that entropy should be defined as the probability of the tendency to move forward towards the most probable macro-state over time (Goldstein & Lebowitz, 2004; Northoff, 2016). Therefore, it is suggested here that entropy vs. free-energy, in this sense, may be described as the molecular events defining the times of a developmental "jump", gestational age-dependent.

Pathology, either physical or mental or both, reflects deviations from very early clocks. In this book three options for deviations from the ontogenetic clock are shown: delayed progress as in the case of fetal growth retardation, acceleration as in the case of accelerated fetal maturity and growth (e.g., thyrotoxicosis, exposure to substance, or pharmacological treatments), or traumatic injury to the clocks for cessation. However, healthy clockwise development triggers the next acquisition towards a new system, which in turn implies numerous feedback mechanisms between the available system and the successive developing one before a new peaking step response, a "jump", can occur. It is further shown in this book that these feedback mechanisms are responsible for the combinatorial pattern of signals between one step response and the following one. Thus, according to this view, the

Chapter 2
The Clocks of Genome Activation: On the Time Designs That Turn Ionic Pathways into the Birth of the First Neuron

Abstract This chapter outlines the clocks of genome activation from the early stage of mobile Ca^{2+} fluxes in the sperm eliciting stable Ca^{2+} currents in the fertilized egg through polarization for the mitotic and postmitotic events creating the apical domain. This chapter also emphasizes the role of clock genes in dictating the times of molecule allocation for protection of the mitochondria resulting from their very early ultradian oscillations, their modulatory interplay with Ca^{2+} currents, their effects on transcriptomic processes of other genes, and their capacity to activate electron transporters for the conversion of unstable molecules to stable ones. The chapter integrates recent data on the design of clock genes as shown in the impact of their ultradian oscillations on chromatin, RohA, and F-actin and their enzymatic cascades, which are the early molecular events for apical domain enrichment and the inner cell population of embryonic development. The chapter continues showing the pathways developing from the early molecular processes of genome activation through the birth of the first neuron, the formation of the neural tube, and its closure. As such, this chapter emphasizes the crucial early role of clock genes in leading towards neural developmental progression before their shift to circadian rhythms upon the maturation of the suprachiasmatic nucleus at mid-gestation. It is concluded that the molecular ultradian oscillations of clock genes during embryonic phases result in time-specific electrical fields when they interact with Ca^{2+} currents. As such, clock genes turn to be an originating source for elicitation, followed by suppression periods, of stereotyped short behavioral cycles, typical in embryos before the gestational timing of the neural tube closure.

Several pivotal processes affect brain development, such as cell differentiation, proliferation, neuralization, migration, and myelination, but what starts the clockwise process of transforming one fertilized egg, a zygote, into a human embryo and then a fetus and a newborn? According to the state of the art, allocation and positioning of cells charges them with the notion of time in a manner that structure and function are combined. It is still unclear what charges early-born cells with the notion of time

S. Goldstein Ferber, *Very Early Clocks*,
https://doi.org/10.1007/978-3-031-89603-3_2

and how these charges are transferred to afford cell localization and specification. What is the pacemaker for very early molecule transport and allocation? In other words, what types of molecules are involved in genome activation?

There is large agreement that development starts with genome activation, but little is known on the clocks of genome activation and the molecular processes that afford it. There is also a traditionally accepted view that early development, particularly healthy fetal behavioral development, depends on the timely emergence of neural networks from the time of neural tube formation and its closure onwards. However, how does a fertilized egg progress to the birth of the first neuron? What is the originating source of very early embryonic stereotyped behaviors occurring before neuralization that can be used as an explanation?

The integration of recent data makes it possible to elucidate the role of clock genes and their interplay with Ca^{2+} in dictating the times of the genome's activation as well as the molecular processes leading to the time when the first neuron is born. The aim of this chapter is to show the molecular events that activate the clock genes' control of the timewise development from the point of Ca^{2+} currents that polarize the sperm and then the zygote up to the point of the first neuron's birth and later during the formation of the neural tube and its closure. Also, this chapter will provide a potential explanation for the originating source of very early embryonic stereotyped behaviors detected in recent years by 3D and 4D ultrasonic recordings.

Mammalian life, initiated at the zygote stage, forms by the fusion of an oocyte and sperm. This fusion generates eight connected blastomeres which are totipotent. Following hours only this eight-cell stage undergoes compaction, in which individual cells become tightly tied at the membranes and become specified for the species. These cells eventually differentiate into the extraembryonic trophectoderm and the pluripotent inner cell mass.

Prior to the implantation of the fertilized egg, the apicobasal domain cell polarity changes from totipotency to pluripotency inducing the segregation clock and cell differentiation to construct the trophectoderm. Genome activation occurs at the four-cell stage and the onset of pluripotency at eight-cell stage (Zhu et al., 2020).

Totipotent stem cells can divide into all cell types in an organism. A totipotent cell has the potential to divide until it creates an entire, complete organism. Pluripotent stem cells can divide into most, or all, cell types in an organism but cannot develop into an entire organism on their own; thus, the change from totipotency to pluripotency reflects the time of genome activation and preparation of the molecules to function as a grand preparatory plan for working as a system, to develop the entire species-specific organism, including all its structures, functions, and their combinations.

Recently, polarization has been found to elicit genome activation. From all ion channels, Ca^{2+} has been found to activate early polarization processes, starting from mobile currents in the sperm cells and changing to rhythmic ones in the fertilized egg through modulating clock genes, which in turn modulate Ca^{2+} flux when activated. This chapter will show the molecular progress of this timewise genome activation. It will show that the clocks of genome activation are governed by Ca^{2+} currents which polarize their bidirectional modulation by the early functions of clock genes. These early functions of clock genes are aimed at (1) eliciting gene conversation along the genome and (2) mastering clocks of somitogenesis through their very early ultradian oscillations.

The polarization of the apicobasal domain, enriched by junction proteins (basal) and apical polarity proteins (apical domain), is critical for the segregation of distinct lineages during mammalian development (Zhu & Zernicka-Goetz, 2020). Polarization occurs in two steps: it begins with the apical enrichment of F-actin, which is promoted by clock genes (Ma et al., 2018) when activated by Ca^{2+}. This beginning is followed by the apical accumulation of the PAR complex, which leads to the formation of an apical domain at the 8–16-cell stage. Clock genes are involved in setting the times for apical polarity timeframes by reducing oxidation caused by F-actin and related enzymatic cascades (Ma et al., 2018). This facilitating function of clock genes is afforded by their activation of electron transporters and cotransporters to convert unusable molecules such as H_2O_2 to stable H_2O and their dynamics of on-off switches on the vector of time (Hansen et al., 2020).

It has been found that the embryonic polarization clock reflects the onset of zygotic genome activation, with three factors that trigger polarization involving Tfap2c, Tead4, and RhoA: (1) the timing of transcription factor AP-2 gamma (Tfap2c) and TEA domain transcription factor 4 (Tead4) expression in the presence of activation of a Ras homolog family member, (2) the timing of RhoA elicitation of polarization and cell fate specification as well as morphogenesis that follows, and (3) the timing of both Tfap2c and Tead4 triggering of actin regulators. The actin expression controls the recruitment of apical proteins on the membrane, whereas RhoA regulates their lateral mobility, providing the origination of the apical domain (Zhu et al., 2020).

Specifically, Zhu et al. (2020) found in mice, an accepted translational model for humans, that Tead4, a transcriptional enhancing factor (a protein) which is encoded by the CEF-3 gene, affects the localization of YAP1, a protein which is encoded by the YES-1 regulator gene, before polarization, and yet it is involved in polarization with Tfap2C, another transcription factor encoded by the AP-2 gamma gene. Furthermore, the two transcription factors, Tead4 and Tfap2C, have been shown to work simultaneously during the two- through eight-cell stage of the zygotic progress, both required at the same time interval for polarization and advancing the zygote to the eight-cell stage. RohA was found to be the crucial contributor in addition to Tfap2c and Tead4 as only with this contribution the apical domain was created at the stage between 2 and 4 days after fertilization (Zhu et al., 2020).

In accordance with the view suggested in this chapter, it was found that CLOCK and BMAL1, both clock genes, stabilize and activate RohA by blocking its degradation, while overexpression of these clock genes regulated an active status of RohA preventing inactivation of RohA as a promoter of healthy molecular events. CLOCK and BMAL1 were found to stabilize and upregulate RohA by inhibiting CLU-3-mediated ubiquitination, a three-step enzymatic cascade of ubiquitin transfer onto amino group of lysine residue, and reducing the interaction in the RohA-ROCK-CFL pathway. The latter alerts the dynamics of F-actin formation (Ma et al., 2018) which in turn is crucial for apical enrichment (Zhu et al., 2020).

In detail, apical domain enables expression of transcription factors which initiate the differentiation into the trophectoderm, whereas apolar cells maintain pluripotency. In mammals, this process splits an embryonic inner cell mass (to be developed into the fetus and newborn) from an extraembryonic trophectoderm trajectory (to be developed into the placenta), known as the developmental trajectories of apical-basal polarity. At the eight-cell stage, polarity is acquired and peaks by cap-like protein domains on the apical surface of each cell (Nishioka et al., 2009). Thus, polarized cells become the trophectoderm, which forms the placenta, and apolar cells become the inner cell mass, the founding population of the fetus (Zhu et al., 2020). The first stages of remodeling embryonic chromatin occur following polarization in the absence of ongoing transcription, during the earlier species-specific period of time. When chromatin states are ordered after fertilization, zygotic genome activation is elicited, and embryonic transcripts turn to function for further control of embryonic development (Østrup et al., 2013). Clock genes in this very early stage set the times via their contribution to chromatin organization, suggesting the early role of clock genes in timing the activation of different parts of the genome. Along one gene, there are parts that are transcribed quicker than other parts. Other findings show the clock genes' ultradian differential elicitation of genomic parts towards transcription (Cox & Takahashi, 2019). This suggests that the early genetic clock-wise input is dependent on timed activation of molecules that accelerate the transcription and others that stop it. Clock genes were shown to accelerate or attenuate other gene transcriptions at this very early stage of development. Recent findings show that clock genes regulate large networks of transcription of other genes by both direct transcriptional activation or repression and recruitment of proteins that modify chromatin to various transcriptional functions (Cox & Takahashi, 2019).

These very early clocks of genome activation require ion transporters, the resulting voltage gradients and electric fields during embryonic development and regeneration before neuralization (McMillen et al., 2020). In animal studies, the genes that control the expression and location of ion fluxes and pumps during early embryonic life have been identified (Rutenberg et al., 2002). However, this process is also complex, suggesting bidirectional impacts flowing from gene control of voltage impacts to ion channels' impact on these expressions and suppression of gene developmental progress over time. Recently, findings support a more specific complex interaction, the one that was found between the Ca^{2+} channels and clock genes (Cavieres-Lepe & Ewer, 2021). How do the clock genes CLOCK/BMAL1, PERIOD ortholog, and Cry1, Cry2 repressor proteins interact with Ca^{2+} fluxes? The clock genes as negative regulators heterodimerize and then translocate into the nucleus repressing there their own gene transcription by interacting directly with CLOCK and BMAL1 (Patke et al., 2017). Contrasting results on effects of Bmal1, show increasing proliferation in early stem cells in Per2 KO mice on the one hand and abolishing of proliferation in Bmal1 KO mice on the other hand (Weger et al., 2017). This suggests that the interplay between the specific genes' clocks dictates the time of proliferation by balancing upregulation and downregulation of the time and rate of proliferation to avoid extreme conditions, either too fast or too slow.

The Ca^{2+} channels are characterized by rapid opening in response to small membrane depolarization. It has been measured that each Ca^{2+} transient lasts for a 33-second period. Also, a lag of about 5h was found when the CAV3 function was required between 16- and 21-cell stage. Regarding calcium channels, the CaV3.2 T-type calcium channel has been identified as crucial during oocyte maturation and following fertilization. Studies in mouse models have shown that CaV3.2 mediates calcium entry necessary for proper calcium store accumulation during oocyte maturation and supports sustained calcium oscillations after fertilization, which are vital for egg activation and early embryonic development. While direct interactions between Cav3 and CaV3.2 in zygotes have not been extensively documented, both are individually important in regulating calcium dynamics during early development. Disruptions in calcium signaling during these stages can lead to developmental abnormalities, highlighting the critical nature of precise calcium regulation in embryogenesis (Bernhardt et al., 2015)

Earlier studies showed that various Ca^{2+} currents were found to control the segmentation clock, a time-controlled process which is termed to describe cell differentiation for the formation of the apical domain (Whitaker, 2006). However, a more complex clock is shown in the crosstalk between clock genes and Ca^{2+} currents, both designed to control on-off switches of divisions by triggering the onset and rate of segmentation at a given period of time in parallel oscillatory sequences.

In these processes, the role of clock genes' function is crucial as they were shown to produce and control electron transporters and cotransporters with antioxidative sensitivity to the allostatic load produced by Ca^{2+} ions (Hansen et al., 2020). The excitatory feedforward manner of the Ca^{2+} clockwise function would be best described as preprogrammed and intrinsically organized on temporal pathways of allostatic load required for progress in healthy gestation controlled by the oscillations of clock genes.

In sum, through eliciting genome activation and creating cascades of transcriptomic progress, mobile Ca^{2+} currents in the sperm initiate fertilization, and later onset of stable Ca^{2+} rhythmic currents initiates zygote formation. Time-sensitive genome activation, partly rapid, while it is slower in other parts of the genome, occurs through the onset of polarization which in turn is responsive to elicitation of the clock genes' time oscillations upregulating and downregulating the transcriptome via bidirectional voltage-gating communicative events with Ca^{2+}. However, how do these processes with their on-off characteristic turn to be the underlying physiological basis of the early occurrence and later elimination of stereotyped short behavioral actions of the embryo?

As shown, following introduction into the egg, the mobile Ca^{2+} waves stimulate the production of InsP3, which in turn spreads across the egg causing the first Ca^{2+} increase. A stable Ca^{2+} wave pacemaker is then formed at the following implantation. Once the zygote is formed, Ca^{2+} acts as the clockwise site originating repetitive Ca^{2+} waves for further cell differentiation through clock genes' ultradian oscillations for advancing differentiation. Identification of early expression patterns by clock genes that modulate ion flux and cell membrane voltage emphasizes the role of calcium transients that are generated by the L-type Ca^{2+} channels during blastula and gastrula stages prior to morphologic differentiation of the nervous system (i.e., the neural tube) and before the emergence of the first neural cell (Webb & Miller, 2006).

The modulation of Ca^{2+} by clock genes determines the time of the first neuron cell's birth up to the time that neural tube closure affects the number of cells born (Cavieres-Lepe & Ewer, 2021). As these clock genes were found to function in early epithelial cells following proliferation (day 4 of stem cells) (Thakur et al., 2020), the process of neuroepithelial cells renewing themselves is shown to be under the time control of these clock genes. Neuroepithelial cells renew themselves while undergoing interkinetic nuclear movements along the apicobasal axis during the cell cycle, producing postmitotic cells that function as newborn neurons. The cooperation of clock genes with the other time signal, Ca^{2+}, is shown in neuroepithelial cells that exhibit a robust increase in nucleoplasm Ca^{2+} in response to G protein-coupled receptor activation when the nucleus is located in the basal region of the cell. This Ca^{2+} rise is caused by the release of Ca^{2+} from intracellular Ca^{2+} stores, reinforcing Ca^{2+} entry from the extracellular space. By exiting this cell cycle, depolarization occurs. The cells lose the apical process where gap junctions are located, causing the postmitotic cells to be charged with a high input resistance, which in turn allows them to be readily depolarized. Thus, mammalian Ca^{2+} channels and transporters (e.g., *ryanodine receptor, ip3r, serca,* and *L*- and *T*-type Ca^{2+} channels) as well as Ca^{2+}-binding proteins (e.g., *camk, cask*, and *calcineurin*) show rhythmic expression and are both modulated and modulating clock gene expression (Cavieres-Lepe & Ewer, 2021). These processes give birth to the first neuronal cells and the onset of the timewise neural tube formation. These very early clocks of genome activation guide also the timely closure of the neural tube (Nikolopoulou et al., 2017).

Taken together, Ca^{2+} fluxes function as an excitatory source that sets time windows and is responsible for polarization periods of time while moving the zygote across differentiation and proliferation phases in a feedforward manner to generate the neural basis for central nervous system development. In accordance with the view presented in this chapter, following somitogenesis, cell-autonomous cycling of Ca^{2+} transients modulates and is modulated by cell-autonomous clock genes' ultradian oscillations for the neural tube development and closure. This process occurs via triggering in-concert asynchronous pulses of actin network to drive apical contraction (Christodoulou & Skourides, 2015). Recently, clock genes have been reported to oscillate in an ultradian manner in the same period of time as Hes7 that is known to control the segmentation clock (Anderson et al., 2020), while the 24-hour cyclicity of fetal clock genes is suppressed until the suprachiasmatic nucleus matures at mid-gestation.

Indeed, although until recently clock genes were considered to be active just from the maturation of the suprachiasmatic nucleus from mid-gestation onwards and the embryo was considered to rely on maternal clocks only (Astiz & Oster, 2021), clock genes, in their bidirectional Ca^{2+} co-modulatory pathways, provide a vector of polarization in time and space earlier, during zygotic and embryonic stages. Through transforming unstable molecules to stable ones, clock genes reduce oxidation, control level of cell oxidation by other proteins, and afford free energy production by available mitochondria (Ferber et al., 2023).

Clock genes function similarly to Ca^{2+} currents via on-off switches of promotion and suppression of molecular events. By these molecular functions, the coordination of clock genes and stable Ca^{2+} fluxes set the time of transcription for other genes and dictate it by very early ultradian rhythms of apical formation and the following onset of transcriptomic processes for neural development.

Beyond their role in differentiation and proliferation, the clock genes' functionality in suppressing and promoting the expression of other genes and their interplay is crucial for designing the time and rate of neuralization (Preitner et al., 2002; Weger et al., 2017). The ultradian manner of their function shows their crucial role within this very early fast-growing organism, excitatory molecular environment, and a high rate of allostatic load. The clock genes' role in dictating resting periods too in their ultradian oscillations shows their importance for healthy embryonic development by regulating the avoidance from unnecessary excitotoxicity and apoptosis in a developmental period that consists of high rates of allostatic load.

Electrical activity and neurotransmitter signaling have been shown to regulate nervous system development even before synapses are formed (Spitzer, 2006). Long after neuralization occurs, neurotransmitter signaling affects neural progenitor proliferation, both of which are cellular processes required for the neural tube formation. However, the control of clockwise development has been exemplified as early as during very early embryonic life and before the first neural cell develops. Recent results in animals show that neural plate cells, which are the basis for the formation, folding, and closure of the neural tube in the embryo, exhibit Ca^{2+} transients which are partially mediated by glutamate through ionotropic NMDA (Arafa & Emara, 2020). This signaling results in regulating proliferation and migration, while any deviation from the optimal range of time is forming a constant status of proliferation which subsequently produces neural tube defects, which can have serious neurological consequences at birth. Thus, increase in proliferative cells occurs when the excitation of glutamate is inhibited and this is accompanied by impaired oriented cell migration. On the other hand, the excitatory impact of glutamate on clock genes and their resulting reaction to disrupt timing of metabolic preprogramed events shows that the interplay between glutamate, NMDA and clock genes is crucial for healthy neuralization by means of coordinated activation of this interplay as early as the differentiation and proliferation stages. This coordinated interplay is maintained through feedback and feedforward mechanisms to keep an optimal range of Ca^{2+} fluxes which also change to be age-appropriate in healthy gestation.

I therefore suggest that the optimal range of time is governed by the crosstalk of clock genes' ultradian oscillations and stable Ca^{2+} currents to determine the developmental timing and action time ranges as early as during the embryonic life throughout neuralization. I further suggest that the waves of this crosstalk function as a clock to avoid faults within the molecular cascade it triggers, such as removing inhibitory signaling including ionotropic Mg^{+} or hypoactivation of NMDA signaling. These waves thus function to define optimal ranges of time to avoid too long or too short periods of polarization and depolarization resulting from abnormal Ca^{2+} dynamics.

The clockwise modulation by Ca^{2+} ions and their bidirectional pathways with clock genes is further important for later stages of fetal development for the healthy formation of the central nervous system. Previous results in animals show the necessity of Ca^{2+} currents for neural tube closure on a temporal pathway (Suzuki et al., 2017).

To further explain conflicting results that EphrinA promoted neural tube closure in mice while others found that this type of ligand leads to neurological defects in other animals (Rasool et al., 2024), I suggest that beyond the difference between species in the different studies, the regulation of critical molecular events for healthy development of the early brain structure and function, the neural tube, is determined by the optimal range of time which is reactive to depolarization shaped by Ca^{2+} current bidirectional modulation of ultradian clock genes' oscillations. Thus, the conflicting results may be explained by different time windows that were examined in the different studies.

Accordingly, I suggest that any deviation in time from optimal ranges of genetic progenitors' impact and their parallel processes, including delays or accelerations of balances in their molecular interactions, may result in DNA insults followed by hyper- or hypo-proliferation, even before the stage of implantation. This points to the crucial role of the clockwise transformation of the genetic input as a very early clock of development.

Recently, others showed that ion channels are detectable in the embryo before the neural tube is developed (Goyal et al., 2020). This is leading to the thought that electrical pulses are involved in the early stages of the biological clock of the genetic input. Indeed, before fertilization, sperm are guided to the egg by a gradient of chemical attractants—a process called chemotaxis. The binding of the chemoattractant to receptors on the surface of the flagellum triggers a cascade of signaling events that eventually lead to an influx of Ca^{2+} ions. Based on these Ca^{2+} surges, which control the waveform of the flagellar beat, sperm adjust their swimming path toward the egg (Alvarez et al., 2014). Additionally, a variety of $InsP_3$-dependent Ca^{2+} signals coded in time, amplitude, and space occur during the process of cardiac cell differentiation and heart development. Studies performed in both embryos and embryonic stem cell differentiating in cardiomyocytes have uncovered that Ca^{2+} regulates multiple steps of these biological processes in vertebrates. These include secretion of cardiogenic factors, cardiac transcriptional cascades and in turn gene expression, myofibrillogenesis, and initiation of embryonic pacemaker activity. Evidence is accumulating to foresee Ca^{2+} as a major second messenger in directing the fate of stem cells and patterning the heart in the embryo (Pucéat & Jaconi, 2005).

Beyond the suggested important role of Ca^{2+} for the creation of the neural tube and the heart very early in gestation, it has been found that the ion channels in general are developed earlier than the neural tube. This ion flux is also likely to have an important role in establishing developmental polarity (Rutenberg et al., 2002; Zhu et al., 2020). However, as noted earlier in this chapter, these crucial roles of Ca^{2+}

currents are dependent of the time designs created very early by the oscillations of clock genes (Cox & Takahashi, 2019).

In sum, the prominent roles of clock genes in genome timely activation, progress of other genes to transcriptional pathways, and timely neuronal birth are shown in their clockwise early functions through their control of molecular events involving F-actin, chromatin, and RohA for the apical domain's very early construction and for the following molecular events required for the birth of the first neuron, the neural tube formation, and its closure. Their capacities for on-off switches and conversion of unstable molecules to stable ones through activation of electron transporters provide clock genes with the underlying design for dictating times and protecting the mitochondria during the mitotic and postmitotic molecular events they elicit.

It is concluded that the molecular ultradian oscillations of clock genes during embryonic phases result in time-specific electrical fields when they interact with Ca^{2+} currents, thus turning to be an originating source for elicitation, followed by suppression periods, of very early stereotyped short behavioral cycles. These very early behavioral time-specific cycles were detected in recent years by 3D, 4D ultrasonic recordings (Donovan et al., 2020; Einspieler et al., 2021) of embryos who were younger than the gestational timing of the neural tube closure. These very early behaviors reflect the process of genome activation, as shown in embryonic early movements, rather than showing an early presentation of any preparation to human interaction. Early preparation to human interaction needs to progress to later stages of neural maturation. However, the prominence of genome activation clocks and their behavioral output for the development of later human behavior is going beyond their coding the transcriptions and is shown in their time-sensitive contribution to the birth of the first neuron and the origination and closure of the neutral tube.

References

Alvarez, L., Friedrich, B. M., Gompper, G., & Kaupp, U. B. (2014). The computational sperm cell. *Trends in Cell Biology, 24*(3), 198–207. https://doi.org/10.1016/j.tcb.2013.10.004

Anderson, M. J., Magidson, V., Kageyama, R., & Lewandoski, M. (2020). Fgf4 maintains Hes7 levels critical for normal somite segmentation clock function. *eLife, 9*, e55608. https://doi.org/10.7554/eLife.55608

Arafa, K., & Emara, M. (2020). Insights about circadian clock and molecular pathogenesis in Gliomas. *Frontiers in Oncology, 10*. https://doi.org/10.3389/fonc.2020.00199

Astiz, M., & Oster, H. (2021). Feto-maternal crosstalk in the development of the circadian clock system. *Frontiers in Neuroscience, 14*. https://doi.org/10.3389/fnins.2020.631687

Bernhardt, M. L., Zhang, Y., Erxleben, C. F., Padilla-Banks, E., McDonough, C. E., Miao, Y.-L., Armstrong, D. L., & Williams, C. J. (2015). CaV3.2 T-type channels mediate Ca2+ entry during oocyte maturation and following fertilization. *Journal of Cell Science, jcs.180026*, 4442. https://doi.org/10.1242/jcs.180026

Cavieres-Lepe, J., & Ewer, J. (2021). Reciprocal relationship between calcium signaling and circadian clocks: implications for calcium homeostasis, clock function, and therapeutics. *Frontiers in Molecular Neuroscience, 14*. https://doi.org/10.3389/fnmol.2021.666673

Christodoulou, N., & Skourides, P. A. (2015). Cell-autonomous Ca2+ flashes elicit pulsed contractions of an apical actin network to drive apical constriction during neural tube closure. *Cell Reports, 13*(10), 2189–2202. https://doi.org/10.1016/j.celrep.2015.11.017

Cox, K. H., & Takahashi, J. S. (2019). Circadian clock genes and the transcriptional architecture of the clock mechanism. *J Mol Endocrinol, 63*, R93. https://doi.org/10.1530/JME-19-0153

Davy, A., & Soriano, P. (2005). Ephrin signaling in vivo: Look both ways. *Developmental dynamics: An Official Publication of the American Association of Anatomists, 232*(1), 1–10.

Donovan, T., Dunn, K., Penman, A., Young, R. J., & Reid, V. M. (2020). Fetal eye movements in response to a visual stimulus. *Brain and Behavior, 10*(8), e01676. https://doi.org/10.1002/brb3.1676

Einspieler, C., Prayer, D., & Marschik, P. B. (2021). Fetal movements: The origin of human behaviour. *Developmental Medicine & Child Neurology, 63*(10), 1142–1148. https://doi.org/10.1111/dmcn.14918

Ferber, S. G., Weller, A., & Soreq, H. (2023). Control systems theory revisited: New insights on the brain clocks of time-to-action. *Frontiers in Neuroscience, 17*. https://doi.org/10.3389/fnins.2023.1171765

Goyal, R., Spencer, K. A., & Borodinsky, L. N. (2020). From neural tube formation through the differentiation of spinal cord neurons: Ion channels in action during neural development. *Frontiers in Molecular Neuroscience, 13*, 62.

Hansen, J. M., Jones, D. P., & Harris, C. (2020). The Redox theory of development. *Antioxidants & Redox Signaling, 32*(10), 715–740. https://doi.org/10.1089/ars.2019.7976

Ma, T., Zhang, Z., Lu, Y., Zhang, Y., Tao, D., Liu, Y., & Ma, Y. (2018). CLOCK and BMAL1 stabilize and activate RHOA to promote F-actin formation in cancer cells. *Experimental & Molecular Medicine, 50*(10), 1–15. https://doi.org/10.1038/s12276-018-0156-4

McMillen, P., Novak, R., & Levin, M. (2020). Toward decoding bioelectric events in *Xenopus* embryogenesis: New methodology for tracking interplay between calcium and resting potentials *In Vivo*. *Journal of Molecular Biology, 432*(2), 605–620. https://doi.org/10.1016/j.jmb.2019.10.029

Nikolopoulou, E., Galea, G. L., Rolo, A., Greene, N. D. E., & Copp, A. J. (2017). Neural tube closure: Cellular, molecular and biomechanical mechanisms. *Development, 144*(4), 552–566. https://doi.org/10.1242/dev.145904

Nishioka, N., Inoue, K., Adachi, K., Kiyonari, H., Ota, M., Ralston, A., Yabuta, N., Hirahara, S., Stephenson, R. O., Ogonuki, N., Makita, R., Kurihara, H., Morin-Kensicki, E. M., Nojima, H., Rossant, J., Nakao, K., Niwa, H., & Sasaki, H. (2009). The Hippo signaling pathway components lats and yap pattern Tead4 activity to distinguish mouse trophectoderm from inner cell mass. *Developmental Cell, 16*(3), 398–410. https://doi.org/10.1016/j.devcel.2009.02.003

Østrup, O., Andersen, I. S., & Collas, P. (2013). Chromatin-linked determinants of zygotic genome activation. *Cellular and Molecular Life Sciences, 70*(8), 1425–1437. https://doi.org/10.1007/s00018-012-1143-x

Patke, A., Murphy, P. J., Onat, O. E., Krieger, A. C., Özçelik, T., Campbell, S. S., & Young, M. W. (2017). Mutation of the human circadian clock gene CRY1 in familial delayed sleep phase disorder. *Cell, 169*(2), 203–215.e13. https://doi.org/10.1016/j.cell.2017.03.027

Preitner, N., Damiola, F., Luis-Lopez-Molina, Z., & J., Duboule, D., Albrecht, U., & Schibler, U. (2002). The Orphan nuclear receptor REV-ERBα controls circadian transcription within the positive limb of the mammalian circadian oscillator. *Cell, 110*(2), 251–260. https://doi.org/10.1016/S0092-8674(02)00825-5

Pucéat, M., & Jaconi, M. (2005). Ca2+ signalling in cardiogenesis. *Cell Calcium, 38*(3), 383–389. https://doi.org/10.1016/j.ceca.2005.06.016

Rasool, D., & Jahani-Asl, A. (2024). Master regulators of neurogenesis: The dynamic roles of Ephrin receptors across diverse cellular niches. *Translational Psychiatry, 14*, 462. https://doi.org/10.1038/s41398-024-03168-4

Rutenberg, J., Cheng, S.-M., & Levin, M. (2002). Early embryonic expression of ion channels and pumps in chick and Xenopus development. *Developmental Dynamics, 225*(4), 469–484. https://doi.org/10.1002/dvdy.10180

Spitzer, N. C. (2006). Electrical activity in early neuronal development. *Nature, 444*(7120), 707–712.

Suzuki, M., Sato, M., Koyama, H., Hara, Y., Hayashi, K., Yasue, N., ... & Ueno, N. (2017). Distinct intracellular Ca^{2+} dynamics regulate apical constriction and differentially contribute to neural tube closure. *Development, 144*(7), 1307–1316.

Thakur, S., Storewala, P., Basak, U., Jalan, N., & Pethe, P. (2020). Clocking the circadian genes in human embryonic stem cells. *Stem Cell Investigation, 7*, 9. https://doi.org/10.21037/sci-2020-014

Webb, S. E., & Miller, A. L. (2006). Ca2+ signaling and early embryonic patterning during the Blastula and Gastrula Periods of Zebrafish and *Xenopus* development. *Biochimica et Biophysica Acta (BBA) – Molecular Cell Research, 1763*(11), 1192–1208. https://doi.org/10.1016/j.bbamcr.2006.08.004

Weger, M., Diotel, N., Dorsemans, A.-C., Dickmeis, T., & Weger, B. D. (2017). Stem cells and the circadian clock. *Developmental Biology, 431*(2), 111–123. https://doi.org/10.1016/j.ydbio.2017.09.012

Whitaker, M. (2006). Calcium at fertilization and in early development. *Physiological Reviews, 86*(1), 25–88. https://doi.org/10.1152/physrev.00023.2005

Zhu, M., & Zernicka-Goetz, M. (2020). Building an apical domain in the early mouse embryo: Lessons, challenges and perspectives. *Current Opinion in Cell Biology, 62*, 144–149. https://doi.org/10.1016/j.ceb.2019.11.005

Zhu, M., Cornwall-Scoones, J., Wang, P., Handford, C. E., Na, J., Thomson, M., & Zernicka-Goetz, M. (2020). Developmental clock and mechanism of de novo polarization of the mouse embryo. *Science, 370*(6522), eabd2703. https://doi.org/10.1126/science.abd2703

Chapter 3
The Clocks of DNA Modifications: On Non-linear Coordination of DNA Translations and Repair

Abstract This chapter presents the unique pattern of very early translational and transcriptomic clocks following genome activation in the zygote. This chapter shows that these clocks consist of numerous tissue-specific clocks coordinated in time, while functioning both acutely and over long periods of time. This chapter also shows the clocks of DNA modification and repair as a sum of inverse U-shaped curves of methylations during acute, very early embryonic phases, which peaks around mid-gestation. This chapter further shows data that contradicts accepted views on an existing negative correlation between the level of methylations and the level of receptor availability. This chapter integrates data showing that in healthy gestation an optimal range of methylation level is required at any given time and that this range narrows as gestation progresses and postnatally. This chapter also argues that DNA modifications regulate gene expression with the capacity of DNA insult repair, including the reactivity to maternal immune activation due to inflammation or exposure to pathogens, especially during early embryonic phases. It is shown that the development of neurobehavioral disorders is not obligatory due to such DNA insults. Finally, this chapter presents DNA modifications as tissue-specific and time-specific, while their peaking time parallels the time of emergence of very early human capacities at mid-gestation, upon reaching the borderline of viability. It is concluded that the timing of coordination between all tissue-specific transcriptomic clocks combining, upregulation of some genes with downregulation of others, is under the governance of clock genes that were found to exert an overarching architecture that affects the entire genome's transcription in a time-sensitive manner.

A mechanism bridging genome activation and the emergence of human behavior is apparent in the developing structure and function of neural networks in the embryo and fetus. The mechanism translating the genetic code represents this bridging process between the genome activation in the zygote and the later phenotypes of embryonic and fetal behavior. How does this mechanism work in healthy gestation and within a healthy developing embryo, fetus, and newborn? How does this mechanism coordinate its clocks along gestation, and what are its peaking time points

S. Goldstein Ferber, *Very Early Clocks*,
https://doi.org/10.1007/978-3-031-89603-3_3

along gestation? What are the data showing that this mechanism is crucially involved in healthy and timely development of mental capacities? What are the data showing that the clocks of this translating mechanism represent the major component of contribution to healthy behavioral development beyond the DNA translations and epigenetic modifications per se? Is this mechanism also capable of repairing DNA insults during early development, and are these potential repairs time-limited?

This chapter aims to outline the answers to these questions. Methylations on genes denote a bridge between genes and expressed phenotype for fetal and neonatal psychobiological development. DNA methylations (DNAm) are chemical modifications of DNA that may be retained as cells divide to make more cells. When found on DNA, methylation can alter gene expression. In this process, chemical tags called methyl groups attach to a particular location within DNA where they turn a gene on or off (or make it easier or harder to translate the gene's message), thereby regulating the production of proteins that the gene encodes (Phillips, 2008). Evidence shown here supports the complex dynamic epigenetic processes in the regulation of transcriptional plasticity within the developing brain and their implications for neurodevelopment, including memory formation, executive functions, goal-directed behaviors, and emotional interpersonal attachment (Wei et al., 2016).

The zygote is endowed with the genetic code diploid from two sets of genes, one from the female egg and the other from the mail sperm. Methylations in the early stage of the embryonic life represent the turn "on" and off" of these genes for further development of the unique individual, the fetus, the newborn, the child, and the adult. At this early stage, one may think that the development of the embryo into the fetus depends solely on gene expression. Enzymes critically associated with changes in DNAm are as follows: DNA methyltransferases and ten-eleven translocation methylcytosine dioxygenases. Epigenetic mechanisms also include posttranslational histone modifications, microRNA (miRNA), long-coding RNA (lncRNA), and noncoding RNA, besides DNAms. miRNAs, subtypes of small, noncoding RNA, 21–25 nucleotides, are capable of base pairing with messenger RNA (mRNA) to modulate gene expression during development and differentiation. By suppressing their expression in a sequence-specific manner, they afford preprogramming of embryonic and fetal development. miRNA can cross the placental barrier, and their regulation may occur in maternal, placental, or fetal tissues. lncRNA are molecules, which can fold into a complex secondary folded structure as containers of miRNA, thus providing a platform for protein isoforms to regulate gene expression and DNA repair as gestation and cellular differentiation progress (Goyal et al., 2019). tRNA-derived small fragments (tRFs) and tRNA halves regulate gene expression, and protein translation, change dramatically in the labyrinth zone along embryonic development, and are tissue-specific spread throughout the embryonic organs, the placenta, and the labyrinth zone located between the spaces of the mother and the spaces of fetus. Research shows acute changes in the levels of specific tRFs and microRNAs which were identified 3–6 h after maternal immune activation, caused by exposure to pathogens or inflammation, and suppressed following the few hours of an acute phase. The complex of tRFs shows that a group of 5' tRNA halves is downregulated by maternal immune activation, whereas a group of 18-nucleotide

Sata, F., Yamada, H., Yamada, A., Kato, E. H., Kataoka, S., Saijo, Y., Kondo, T., Tamaki, J., Minakami, H., & Kishi, R. (2003b). A polymorphism in the CYP17 gene relates to the risk of recurrent pregnancy loss. *Molecular Human Reproduction, 9*(11), 725–728. https://doi.org/10.1093/molehr/gag089

Scarpulla, R. C. (2006). Nuclear control of respiratory gene expression in mammalian cells. *Journal of Cellular Biochemistry, 97*(4), 673–683. https://doi.org/10.1002/jcb.20743

Sies, H., Berndt, C., & Jones, D. P. (2017). Oxidative Stress. *Annual Review of Biochemistry, 86*, 715–748. https://doi.org/10.1146/annurev-biochem-061516-045037

Tadesse, S., Kidane, D., Guller, S., Luo, T., Norwitz, N. G., Arcuri, F., Toti, P., & Norwitz, E. R. (2014). In vivo and in vitro evidence for placental DNA damage in Preeclampsia. *PLoS One, 9*(1), e86791. https://doi.org/10.1371/journal.pone.0086791

Tempfer, C., Unfried, G., Zeillinger, R., Hefler, L., Nagele, F., & Huber, J. C. (2001). Endothelial nitric oxide synthase gene polymorphism in women with idiopathic recurrent miscarriage. *Human Reproduction, 16*(8), 1644–1647. https://doi.org/10.1093/humrep/16.8.1644

Torres-Cuevas, I., Parra-Llorca, A., Sánchez-Illana, A., Nuñez-Ramiro, A., Kuligowski, J., Cháfer-Pericás, C., Cernada, M., Escobar, J., & Vento, M. (2017). Oxygen and oxidative stress in the perinatal period. *Redox Biology, 12*, 674–681. https://doi.org/10.1016/j.redox.2017.03.011

Welin, A.-K., Sandberg, M., Lindblom, A., Arvidsson, P., Nilsson, U. A., Kjellmer, I., & Mallard, C. (2005). White matter injury following prolonged free radical formation in the 0.65 gestation fetal sheep brain. *Pediatric Research, 58*(1), 100–105. https://doi.org/10.1203/01.PDR.0000163388.04017.26

Wharfe, M. D., Mark, P. J., Wyrwoll, C. S., Smith, J. T., Yap, C., Clarke, M. W., & Waddell, B. J. (2016). Pregnancy-induced adaptations of the central circadian clock and maternal glucocorticoids. *Journal of Endocrinology, 228*(3), 135–147. https://doi.org/10.1530/JOE-15-0405

Zhou, G., Winn, E., Nguyen, D., Kasten, E. P., Petroff, M. G., & Hoffmann, H. M. (2022). Co-alterations of circadian clock gene transcripts in human placenta in preeclampsia. *Scientific Reports, 12*(1), 17856. https://doi.org/10.1038/s41598-022-22507-3

Part II
OUTPUTS—The Rise of Psychological Phenomena

Chapter 6
The Clocks of Self-Regulation: Rhythms of ATPases Generate Behavioral Regulation

Abstract This chapter serves as the opening chapter of Part II in this book, and as such it presents the general view of Part II, describing the origins of early human behavioral clocks as based on the maturation on specific neural networks. Self-regulation can be described as the capacity of the organism to return to balance after mounting arousals. This definition applies to both physiological and behavioral processes. As such it serves as an umbrella for all neural-based behavioral accomplishments from mid-gestation onwards. The clocks of self-regulation presented in this chapter show that behaviors exist even before the neural tube closes, while their self-regulatory characteristics emerge only at mid-gestation. This chapter argues that early behavioral regulation is dependent on the timely electron transport and accurate expression of ATPases. Specifically, it is accepted that generating the human six-layered cerebral cortex, which serves for the elicitation of more regulated behaviors, is dependent on integrating signals from various cell types. This chapter shows that this accomplishment is dependent on the ionic gradients of the cell membrane potential and accurate function of the ATPases. Furthermore, the molecular rhythms that lead firing rates and their timing during embryonic life develop through mid-gestation to afford the elicitation of early fetal self-regulatory behaviors and their return to balance after mounting arousal. This chapter also shows that ATPases events are reversible and shaped by behavior from mid-gestation onwards. This includes feedback mechanisms originating in self-regulatory behaviors and transmitted back to the molecular level. At the time window of mid-gestation, ATPases reduce the dominance of excitation and start to balance it under the control of clock genes which by then gradually shift from ultradian to circadian oscillations. It is concluded that the ATPases function as the early flexible basis of the emerging self-regulated behaviors, while timing of changes in ATPases expression is controlled by clock genes. ATPases thus control the timing of behavioral regulation, covering by that a large umbrella of emerging self-regulatory behaviors from mid-gestation onwards.

What is self-regulation and does this capacity serve as an umbrella for all human behavior? Can this capacity be applied even to very early behaviors? Is self-regulation elicited according to timely trajectories during gestation? What is the

S. Goldstein Ferber, *Very Early Clocks*,
https://doi.org/10.1007/978-3-031-89603-3_6

physiological basis for the development of self-regulation of fetal behaviors? How can the fetus self-regulate early behaviors with its own resources? This chapter aims to answer these questions. Self-regulation can be described as the capacity of the organism to return to balance after mounting arousal (Ferber, 2008). This chapter will expand on my previous work (Goldstein Ferber et al., 2022), by arguing that the emergence of the self-regulation clock precedes other early psychological phenomena, including executive functions and goal-oriented behavior.

In Part I of this book, clock genes were presented as the designers of developmental trajectories on the vector of time including clocks of genome activation, clocks of DNA modifications, clocks of myelination, and clocks of antioxidative defenses. As such, Part I presented the building blocks for the onset of fetal human behavior around mid-gestation. This chapter is the opening chapter for Part II of this book, describing the behavioral outputs with their developmental changes from early embryonic phases to late gestation, focusing on the developmental junction of each type of behavior at mid-gestation, the borderline of viability timing. The clocks of self-regulation are presented as the opening chapter for Part II, emphasizing that the emerging healthy human psychological phenomena require the umbrella of self-regulation capacities beyond specific brain interregional development and the closure of the neural tube. The umbrella of self-regulation clocks will be presented in this chapter with the specific fetal behaviors it embraces and molecular events associated with their onset on the vector of time.

Recent recordings by 3D, 4D ultrasonic techniques show the emergence of spontaneous and reactive fetal behaviors and their developmental patterns along healthy gestation. However, what regulates the change from stereotyped cyclic behaviors of the embryo to more organized stimulus-related behaviors after mid-gestation? Additionally, since these late-gestation behaviors occur as relatively organized though short in time, what is the organismal regulator of the psychological self-regulation functionality? And as this onset of self-regulated behaviors occurs in an age-dependent manner, how and when does this regulator of the overall self-regulation capacities develop in time?

On the cellular level, during early embryonic life, the creation of the Na^+K^+ATPase pump in the neuronal cell determines the firing time and the recovery time of the cell. These intrinsic currents are followed by non-synaptic calcium plateau and synaptic depolarization potentials. This chapter will argue that increasing and selecting neural connectivity are processes controlled by the accumulating coordinated activity of these ion channels to function for the developing early ability of the fetus to behaviorally return to balance after mounting arousal. This chapter will also argue that the onset of relatively regulated ATPases coincides with the elicitation of more organized fetal behaviors. The chapter will end with the argument, still to be investigated, that self-regulatory behaviors function as feedback mechanisms for the molecular level and its developing coordination from the time of their emergence onwards.

The bridge between embryonic stereotyped cyclic behaviors, and the occurrences of partial awareness incidents with the developing self-regulation from mid-gestation onwards, may be viewed evolutionarily as part of pruning, synapse selection, and the emerging trial and error of the human experience aimed at early learning to react adaptively to a future out-uterine stimulus. From the time of mid-gestation onwards, it is suggested here that the mindful self-regulatory behaviors function as feedback mechanisms to their molecular basis, earlier established and regulated, to achieve their further fine-tuning and required signal coordination. This is an accomplishment of the individual organism who exhibits these behaviors through his/her own cell specification and alteration processes as environmentally required and uniquely developed (Ferber et al., 2022).

Regulation of clocks and their temporal sequence in very early development can be understood as self-organization of the organism, starting at the atomic level and the level of ion transporters. One may wonder: what do electron transporters have to do with later developing psychological phenomena and postnatally observed behavioral balance, termed as "self-regulation"? However, generating the human six-layered cerebral cortex is dependent on integrating signals from various cell types coordinating cell migration, proliferation, and differentiation, an accomplishment which in turn is dependent on the ionic gradients of the cell membrane potential and accurate function of the pumps.

Specifically, cell differentiation and specification, including cortical folding and expansion, have been found to be closely affected by ATPases and their transcriptomic function in gene expression, such as ATP1A3, highly expressed in the subset of deep-layer neurons in the subplate (Smith et al., 2021) as well as the de novo gene SP0535 (Qi et al., 2023). Notably, these genes have been found to underlie cognitive and memory functions (Qi et al., 2023) which are capacities that are behavioral and used under the capacity of self-regulation during childhood when facing challenging environmental stimuli. Interestingly, these two types of genes, encoding ATPases, have a fundamental role in regulating the communication between excitatory and inhibitory neurons, i.e., modulation of arousal and return to balance, as expressed in the self-regulation definition.

Beyond their photic reactivity and the onset of fetal circadian rhythms upon the relative maturation of the suprachiasmatic nucleus at mid-gestation, clock genes were found to regulate extracellular ATP (Wang et al., 2023). Specifically, they regulate the $Na^{+}K^{+}$ beta1 subunit expression (Nakashima et al., 2018). These findings suggest that clock genes with their heterodimerizing functions and rhythmic ultradian oscillatory transcriptions are the guardians of time to regulate ATPase chromatin-related molecular events.

To date, several types of ATPases are known, while many others remain still to be discovered: the P-type (mainly $Na^{+}K^{+}$ ATPase) reverse complexes known as F-type ATPases, vacuolate V-type ATPases, and the largest group of ABC-type, referring to dual sequences twice on the same polypeptide chain, implicating two nucleotide binding sites, although the two may be found on the polypeptide chain too. In terms of subunits, F and related V-type ATPases are the most complex, whereas the complete ATPase synthesis/ATPase complex, excluding regulators, contains at least 15 subunit types (Pedersen, 2005). These subunits show an intricate interplay between themselves varying between lower, higher, or no change in transport rates along gestation on the vector of time (Floyd et al., 2017).

To emphasize the role of ion channels such as ATPases to the elevation of timely behavioral reactivity, it is noted that early sources for neurologically related behavioral defects have been termed "channelopathy" (Smith et al., 2021), referring to interferences in the optimal function of the early developing ATPases, i.e., pathogenic mutations in genes including de novo genes emerging from non-encoded chains that encode the ion channels. This role is apparent in ATP synthesis, where in the Na^+K^+ ATPase, 3 Na^+ ions move out of the cell and 2 K^+ ions enter the cell, causing negative cell membrane potential and contributing to the differential concentration gradients of sodium and potassium for the firing rate and resting state of the cell. Developmental tissue-specific differences in isoforms show that ATPases function to be accurately regulated directly (Na^+K^+) or indirectly (Ca^{2+} Cl^-) (Floyd et al., 2017).

Thus, any encoding malformation may cause deformities in the cortical developing structure and their aligned behavioral defects and diseases such as epileptic encephalopathy, hemiplegia of childhood, cerebellar ataxia, and sensorineural hearing loss among other neurological diseases (Brown & García-García, 2018). In general, such variants may lead to deviations from timely contribution of the ATPase pumps to polarization-depolarization conditions ending in excitotoxicity and danger to the life of the neuron as resulting from interferences with the return to balance and resting state. The crucial role of ATPases is also apparent in their capacity to alter specifications of cells, thus changing the optimal preprogrammed manner of embryonic and fetal cortical regulatory development and its aligned behavioral phenotypes. This capacity of ATPases to alter cell specification is suggested as the molecular output of the feedback mechanisms provided to the molecular level by execution of self-regulatory behaviors.

The very early contribution of the ATPases has been recognized in the embryonic transition from morula to blastocyte (Houghton et al., 2003). At this very early stage of development, embryonic life, up to 60% of the oxygen supply to the embryo is used by the ATPase pumps (Houghton et al., 2003) emphasizing the contribution of electron transporters to convergence of oxidative molecules such as H_20_2 into water (Baker & Ebert, 2013; Schafer & Buettner, 1999), and thus to mitochondrial energy supplies for further preprogrammed elicitation of behavioral reactivity. During embryonic life, bioelectric function in brain development required for the propagation of signals implies ion channels and pumps as spatiotemporal regulators of morphogenetic processes which include the preprogrammed specification of cells' function and their alterations in conditions of deviation from the ordered clocks of the pumps, thus affecting the elicitation of behavioral phenotypes. In later stages, especially at neural tube closure, at 28 weeks of gestation, a crucial stage for further resolution of the cortical subplate and further folding of the developing cortex, the clocks of the ATPases turn to be the granular regulatory basis for the emergence of behavioral phenotypes and their goal-oriented reactivity to a given intrauterine stimulus (Qi et al., 2023; Zhu et al., 2020). From mid-gestation onwards, regulated ATPases serve again as the crossroad of intrauterine life from embryonic into fetal signaling and behavioral progress, avoiding excitotoxicity and promoting cortical folding and expansion, beyond membrane growth and synapse selection itself. At this stage of gestation, synapse selection may be shaped by emerging self-regulatory behaviors and their feedback impact on the molecular level.

While it is known for decades that the neural tube develops into the CNS structures in mammals and other species, a crucial crossroad appears at 28 weeks of human gestation, the timing of neural tube closure, pointing to another significant aspect of mid-gestation timing. Interferences with this closure are responsible for 6% of the infants worldwide suffering from congenital malformations, according to the WHO (Isaković et al., 2022). Recent studies show that coordinated polarized locations across the developing neural tube activated on the noncanonical Wnt/planar cell polarity pathways by transcription factors Grh2/3, Pax3, Cdx2, and Zic2 act in concert to adjust closure of neural tube structures, including convergent extension, apical constriction, and interkinetic nuclear migration, thus creating the folds by these induced functions and structures affected by the ATPases.

Therefore, this spatiotemporal process is dependent on a coordinated clock of polarization (Nikolopoulou et al., 2017). As such, it is dependent on calcium fluxes and the Na^+K^+ ATPases (Brown & García-García, 2018). This appears to be the major granular building block responsible for coupled and self-regulatory fetal behaviors which were detected following the time window of mid-gestation (Einspieler et al., 2021). The dependence of neural tube closure in healthy fetal development on coordinated timing of polarization in several predefined locations presents the crucial role of ATPses in creating the structural and functional basis of early self-regulatory human behavior.

Gestation induces specific regulatory changes in expression of at least the Na^+ K^+ ATPase isoforms in the uterus, which peak at mid-gestation in the interplay of its subunits during healthy pregnancy, affecting contractility and preparatory phases for healthy delivery, as shown in animal models (Floyd et al., 2017). Additionally, from mid-gestation onwards, coupling of discriminated sleep states, heart rate, and fetal behaviors, including reflexes and more coordinated behavioral reactivity, was observed close to birth (Ferber et al., 2023a). Thus, this complex achievement of fetal behavioral development is possible only if polarization of neural cells is coordinated and ATPases pumps support a timely interplay between polarized and depolarized neural network locations, to regulate excitation of rapid development during embryonic life as well as arising fluctuations of inhibition for the maturation of excitation-inhibition balance and the resulting behavioral self-regulation by mid-gestation. This may be regarded from an evolutionary perspective as preparation for healthy out-uterine behavioral reaction postnatally.

Thus, the individual's self-regulatory capacities have a granular basis on the molecular level of the ATPase pumps as early as the blastocyte function during embryonic life (Pedersen, 2005). As shown here, the germinal origins of neural network connectivity start even before the timing of the neural tube closure at 28 weeks of gestation reaching a peak towards birth (De Paepe et al., 2014; Isaković et al., 2022; Smith et al., 2021).

As the polarization-depolarization control of ATPases is apparent even before neural tube closure and it is associated with premature behavioral phenotypes, it exemplifies the basis for the rise of regulated behavior as early as the end of the first trimester in a healthy pregnancy. One important example of the early rise of self-regulatory behaviors is the 12 weeks' gestational age onset of a lifelong behavior apparent across species, fetal yawning, coupled frequently with the movement of the hand to mouth. A fetal yawn with wide opening of the mouth lasts 2–8 seconds followed by a quick shut of the mouth (Einspieler et al., 2012). The yawning with hand-to-mouth complex has been mentioned as a fetal behavioral regulator of arousal, oriented as preparation for future communication, since it increases the circulation of cerebrospinal fluid. Additionally, from 14 weeks of gestation onwards, the fetus shows rhythmic occasions of jaw opening followed by amniotic fluid swallowing, indicating drinking of the fluid, evidenced by flow signals in the oro-nasal cavity (de Vries et al., 1982; Einspieler et al., 2012). Fetal sucking is preparatory for out-uterine vital food intake. It increases after mid-gestation towards birth and functions as one of the amniotic fluid volume regulators (Barlow, 2009; Einspieler et al., 2012). Occasionally, the fetus sucks the umbilical cord too. Other examples of the rise of self-regulation, observed later, after mid-gestation and onwards by 3D, 4D ultrasound recording, are facial expressions such as smiling, pouting, and scowling, including coupled eye movements and nonregulated nystagmus (Einspieler et al., 2021). These are behaviors clearly aimed at future human communication, and they are evolutionarily shaped behaviors to attract the caregiver's attention and sympathy.

The sensitive time of mid-gestation has been also identified as the time associated with maturation of sleep states, including the onset of REM sleep and the discrimination between quiet sleep and active sleep. Furthermore, the maturation of sleep states is coupled with heart rate regulation and reduction in beat per minute as well as with behavioral reactions and onset of facial expressions in response to sensory inputs, as shown by 3D ultrasound recordings and laser dot diodes (Donovan et al., 2020). This later achieved accomplishments occur on the expense of random and cyclic general limb and body movements which were observed earlier in gestation (Einspieler et al., 2021). I have previously identified that at this time window of mid-gestation, all the three parameters, heart rate, sleep states, and a complex of motor behaviors (rather than isolated movements), correlate and are coupled (Ferber et al., 2023a), suggesting more mature connectivity between sensorimotor and associative regions and more coherence in the reactivity of fetal behaviors to sensory intrauterine inputs (Donovan et al., 2020).

The significant developmental time window of mid-gestation signals the junction of future healthy cortical development, switching from sensory bombardment to more accurate electrical signaling with further waves of myelination and cell migration within the loop of molecular coordination for the emergence of self-regulatory behaviors. These behaviors, in turn, function within a feedback loop to the molecular level when experience-shaped, a process known to date as pruning, occurs. Additionally, by the time behaviors are coupled with these more accurate currents at mid-gestation, a preprogrammed "stop signal" for the neuronal burst activity is apparent by the voltage-gated K^+ signaling. Thereafter, "senior" older neurons wire together to create networks with other "senior" neurons" (Goldstein Ferber et al.,

2022). As shown elsewhere, this process occurs earlier in "older" structures such as the brainstem and spinal cord and later in "younger" structures such as the cortex.

This observation is validated by findings that fetal coupling between heart rate, sleep states, and behavior at this gestational time window predicts high levels of effortful behavioral control and improved self-regulation during childhood and adolescence (Van den Bergh & Mulder, 2012). Interestingly, at mid-gestation closure of the neural tube meets the penetration of thalamic fiber into the cortical plate, continuing the resolution of the subplate, maturation of sensorimotor regions, and the six-layered cortical early structure. These complex processes occurring at mid-gestation clearly aim at future human communication and are the early basis of fetal self-regulation.

Low self-regulation of preterm infants has been reported repeatedly during the past decades (for review, see Ferber et al., 2022). It is possible that the low capacity for self-regulation of preterm infants, especially those born at mid-gestation, is not only due to early exposure to the extra-uterine environment as thought to date, but also due to lasting embryonic molecular imbalances. One example is that the capacity to self-regulate psychological experience and behavior postnatally, known in the literature as "self-regulation," is shown here to be based on the creation of changing balances during a period of major excitation and rapid development of the embryo, before the interchange of GABA from its early excitatory role to its traditional role of inhibition (Herlenius & Lagercrantz, 2004). Thus, early molecular systems with defects in optimal excitation-inhibition balance would be a fragile factor later in gestation and following birth for the emergence of self-regulation by the human fetus and newborn. Such a fragile molecular basis interferes with the timely feedback mechanisms originating in behavioral inputs and transmitted to the molecular levels. Such faults further increase the chances for the development of neurological diseases and deformed behavioral phenotypes, which may be better described as insults to the clocks of self-regulation, their molecular basis and their effects on accuracy of signaling.

The molecular basis for the developing self-regulation may also show manners for compensation and reversibility to their optimal function (Ferber et al., 2024) with or without pharmacological treatment of the pregnant mother. The zygote starts a complex journey. A close focus on each stage of development will show that this journey is not deterministic, and it may have many fluctuations in its general linear development from fertilization to birth, including natural compensation, shown in redundant backup systems (Ferber, Weller, & Soreq, 2023b). According to the view presented in this chapter, self-regulation clocks occur within the loop of feedback mechanisms to the molecular level. The errors revealed between the behavioral output and the molecular events, which are associated with their timing, bring about the involvement of ATPases in alteration of cell specifications. Defects and pathological conditions may be reversed to optimum due to accumulating occasional balances or by pharmacogenomic therapeutic agents supporting ATPases in managing the challenges of neural tube closure, cortical folding, resolution of the subplate, and conditions of channelopathy through supporting gradual occurrences of excitation-inhibition balance.

It is further argued here that the borderline of viability, early mid-gestation, which is accepted as the gestational age of pain perception appearance upon relative maturation of sensorimotor regions (Kostović & Judaš, 2010; Lagercrantz & Ringstedt, 2001), is a crucial time window of self-regulatory development. Earlier developing progress of fluctuations between neural arousal and return to balance through premature polarization-depolarization variability of neural networks precedes this determining time window at mid-gestation. The development in early mid-gestation has a crucial role in further development of self-regulation during intra- and extrauterine early life and on the level of susceptibility to environmental influences.

To outline the clocks of self-regulation, it is noted that healthy developmental clocks of self-regulation occur only when the molecular basis turns from high levels of excitation, sometimes recognized as excitotoxicity, early in gestation, to fluctuating excitation-inhibition balance as shown in the coinciding transition from isolated fetal movements to complex and stimulus-related behaviors. In this sense, the timely molecular functionality of polarization-depolarization, progressing from rapid development, based on prolonged excitation in the embryo, to the emergence of inhibitory signals and phases of excitation-inhibition balance at mid-gestation, secures the basis of the newborn's capacity to behaviorally return to balance and self-regulate after birth, very early in life.

In sum, the molecular coordination leading excessive firing rate at frequent timings as early as embryonic life affords the elicitation of early fetal self-regulatory behaviors and the return to balance after mounting arousal at mid-gestation. This includes feedback mechanisms originating in self-regulatory behaviors transmitted back to the molecular level in a gestational age-dependent fluctuating manner, not necessarily in a simple linear increase of self-regulation. It is concluded that the ATPases function as the general regulators of the emerging self-regulated behaviors, while timing of changes in ATPase expression is controlled by the ultradian oscillations of clock genes. ATPases thus serve as the regulators of the emerging self-regulatory clocks, covering by that a large umbrella of complex fetal behaviors at mid-gestation and onwards.

References

Baker, C., & Ebert, S. (2013). *Development of aerobic metabolism in utero: Requirement for mitochondrial function during embryonic and foetal periods* (Vol. 2).

Barlow, S. M. (2009). Central pattern generation involved in oral and respiratory control for feeding in the term infant. *Current Opinion in Otolaryngology & Head and Neck Surgery, 17*(3), 187. https://doi.org/10.1097/MOO.0b013e32832b312a

Brown, J. M., & García-García, M. J. (2018). Secretory pathway calcium ATPase 1 (SPCA1) controls mouse neural tube closure by regulating cytoskeletal dynamics. *Development, 145*(19), dev170019. https://doi.org/10.1242/dev.170019

De Paepe, C., Krivega, M., Cauffman, G., Geens, M., & Van de Velde, H. (2014). Totipotency and lineage segregation in the human embryo. *Molecular Human Reproduction, 20*(7), 599–618. https://doi.org/10.1093/molehr/gau027

de Vries, J. I. P., Visser, G. H. A., & Prechtl, H. F. R. (1982). The emergence of fetal behaviour. I. Qualitative aspects. *Early Human Development, 7*(4), 301–322. https://doi.org/10.1016/0378-3782(82)90033-0

Donovan, T., Dunn, K., Penman, A., Young, R. J., & Reid, V. M. (2020). Fetal eye movements in response to a visual stimulus. *Brain and Behavior, 10*(8), e01676. https://doi.org/10.1002/brb3.1676

Einspieler, C., Prayer, D., & Marschik, P. B. (2021). Fetal movements: The origin of human behaviour. *Developmental Medicine & Child Neurology, 63*(10), 1142–1148. https://doi.org/10.1111/dmcn.14918

Einspieler, C., Prayer, D., & Prechtl, H. F. R. (2012). *Fetal behaviour: A neurodevelopmental approach*. Mac Keith Press. https://cir.nii.ac.jp/crid/1130282270595194880

Ferber, S. G. (2008). The concept of coregulation between neurobehavioral subsystems: The logic interplay between excitatory and inhibitory ends. *Behavioral and Brain Sciences, 31*(3), 337–338. https://doi.org/10.1017/S0140525X08004123

Ferber, S. G., Als, H., McAnulty, G., Klinger, G., & Weller, A. (2022). Multi-level hypothalamic neuromodulation of self-regulation and cognition in preterm infants: Towards a control systems model. *Comprehensive Psychoneuroendocrinology, 9*, 100109. https://doi.org/10.1016/j.cpnec.2021.100109

Ferber, S. G., Geva, R., & Weller, A. (2023a). When the mind comes to live inside the body: The ontogeny of the perceptual control clock. *Current Neuropharmacology, 21*(1), 13–21. https://doi.org/10.2174/1570159X20666220411095508

Ferber, S. G., Weller, A., & Soreq, H. (2023b). Control systems theory revisited: New insights on the brain clocks of time-to-action. *Frontiers in Neuroscience, 17*. https://doi.org/10.3389/fnins.2023.1171765

Ferber, S. G., Weller, A., & Soreq, H. (2024). Boltzmann's theorem revisited: Inaccurate time-to-action clocks in affective disorders. *Current Neuropharmacology, 22*(11), 1762–1777. https://doi.org/10.2174/1570159X22666240315100326

Floyd, R. V., Mobasheri, A., & Wray, S. (2017). Gestation changes sodium pump isoform expression, leading to changes in ouabain sensitivity, contractility, and intracellular calcium in rat uterus. *Physiological Reports, 5*(23), e13527. https://doi.org/10.14814/phy2.13527

Goldstein Ferber, S., Weller, A., Ben-Shachar, M., Klinger, G., & Geva, R. (2022). Development of the ontogenetic self-regulation clock. *International Journal of Molecular Sciences, 23*(2), Article 2. https://doi.org/10.3390/ijms23020993

Herlenius, E., & Lagercrantz, H. (2004). Development of neurotransmitter systems during critical periods. *Experimental Neurology, 190*, 8–21. https://doi.org/10.1016/j.expneurol.2004.03.027

Houghton, F. D., Humpherson, P. G., Hawkhead, J. A., Hall, C. J., & Leese, H. J. (2003). Na+, K+, ATPase activity in the human and bovine preimplantation embryo. *Developmental Biology, 263*(2), 360–366. https://doi.org/10.1016/j.ydbio.2003.07.014

Isaković, J., Šimunić, I., Jagečić, D., Hribljan, V., & Mitrečić, D. (2022). Overview of Neural Tube Defects: Gene–Environment Interactions, Preventative Approaches and Future Perspectives. *Biomedicines, 10*(5), Article 5. https://doi.org/10.3390/biomedicines10050965

Kostović, I., & Judaš, M. (2010). The development of the subplate and thalamocortical connections in the human foetal brain. *Acta Paediatrica, 99*(8), 1119–1127. https://doi.org/10.1111/j.1651-2227.2010.01811.x

Lagercrantz, H., & Ringstedt, T. (2001). Organization of the neuronal circuits in the central nervous system during development. *Acta Paediatrica, 90*(7), 707–715. https://doi.org/10.1111/j.1651-2227.2001.tb02792.x

Nakashima, A., Kawamoto, T., Noshiro, M., Ueno, T., Doi, S., Honda, K., Maruhashi, T., Noma, K., Honma, S., Masaki, T., Higashi, Y., & Kato, Y. (2018). Dec1 and CLOCK regulate Na+/K+-ATPase β1 subunit expression and blood pressure. *Hypertension, 72*(3), 746–754. https://doi.org/10.1161/HYPERTENSIONAHA.118.11075

Nikolopoulou, E., Galea, G. L., Rolo, A., Greene, N. D. E., & Copp, A. J. (2017). Neural tube closure: Cellular, molecular and biomechanical mechanisms. *Development, 144*(4), 552–566. https://doi.org/10.1242/dev.145904

Pedersen, P. L. (2005). Transport ATPases: structure, motors, mechanism and medicine: A brief overview. *Journal of Bioenergetics and Biomembranes, 37*(6), 349–357. https://doi.org/10.1007/s10863-005-9470-3

Qi, J., Mo, F., An, N. A., Mi, T., Wang, J., Qi, J.-T., Li, X., Zhang, B., Xia, L., Lu, Y., Sun, G., Wang, X., Li, C.-Y., & Hu, B. (2023). A human-specific de novo gene promotes cortical expansion and folding. *Advanced Science, 10*(7), 2204140. https://doi.org/10.1002/advs.202204140

Schafer, F. Q., & Buettner, G. R. (1999). Singlet oxygen toxicity is cell line-dependent: A study of lipid peroxidation in nine leukemia cell lines. *Photochemistry and Photobiology, 70*(6), 858–867. https://doi.org/10.1111/j.1751-1097.1999.tb08294.x

Smith, R. S., Florio, M., Akula, S. K., Neil, J. E., Wang, Y., Hill, R. S., Goldman, M., Mullally, C. D., Reed, N., Bello-Espinosa, L., Flores-Sarnat, L., Monteiro, F. P., Erasmo, C. B., Pinto e Vairo, F., Morava, E., Barkovich, A. J., Gonzalez-Heydrich, J., Brownstein, C. A., McCarroll, S. A., & Walsh, C. A. (2021). Early role for a Na+,K+-ATPase (ATP1A3) in brain development. *Proceedings of the National Academy of Sciences, 118*(25), e2023333118. https://doi.org/10.1073/pnas.2023333118

Van den Bergh, B. R. H., & Mulder, E. J. H. (2012). Fetal sleep organization: A biological precursor of self-regulation in childhood and adolescence? *Biological Psychology, 89*(3), 584–590. https://doi.org/10.1016/j.biopsycho.2012.01.003

Wang, X., Dong, Y.-T., Hu, X.-M., Zhang, J.-Z., Shi, N.-R., Zuo, Y.-Q., & Wang, X. (2023). The circadian regulation of extracellular ATP. *Purinergic Signalling, 19*(1), 283–295. https://doi.org/10.1007/s11302-022-09881-3

Zhu, M., Cornwall-Scoones, J., Wang, P., Handford, C. E., Na, J., Thomson, M., & Zernicka-Goetz, M. (2020). Developmental clock and mechanism of de novo polarization of the mouse embryo. *Science, 370*(6522), eabd2703. https://doi.org/10.1126/science.abd2703

Chapter 7
The Clocks of Perception: Rhythms of Ionic Gradients Shape Partial Awareness Incidents

Abstract In recent years, changes from early embryonic stereotyped behaviors to stimulus-related fetal behaviors have been recorded following mid-gestation. These include capacities such as habituation and orienting response, which incorporate a transitory quality of awareness. Given these recent data, this chapter shows that more organized fetal behaviors during the last trimester are based on biological components which are known to construct neural wiring. Among these biological components, this chapter notes that following embryonic rapid phases of development NMDA mediated by AMPA serves as an inhibition agent for the GABA-glutamate excitatory interplay. This rise of inhibitory factors at mid-gestation is controlled by Cl^- mobilization of Ca^{2+} cell entries towards more balanced frequencies of neural cell's action potentials as part of the cell's increasing readiness to partner in postsynaptic transmission and wiring. Glial cells are able to define potential errors in wiring even before the maturation of any specific wiring. Later they regulate the osmotic homeostasis in the neural cell through choroid plexus macrophages, thus contributing to the speed and fate of wiring. Recent data suggest that the rhythms of cellular osmosis are controlled by clock genes through their typical posttranslational regulation of protein activity, ion transport, and consolidation of proteome renewal cadences. The timing of changes in the Cl^- dynamics and in the type of clock gene oscillations coincides with the onset of experience-shaped synaptic changes. This implies that by this timing, at mid-gestation, feedback loops exist between the behavioral and the molecular levels, resulting in the onset of coordinated firing of neural networks and corresponding stimulus-responsive behavioral outcomes. Thus, the clocks of perception are described in this chapter as produced by molecular-based sustained wiring through synapse selection. This in turn results in a gradual active collection of germinal transitory quality of awareness peaks, for any given behavior at any given intrauterine compound from mid-gestation onwards. In this sense, very early perceptions occur in instances when molecular events meet morphological readiness to exhibit a behavior and absorb its results at the same time. The expansion of these instances depends on the rhythms of their repetition.

S. Goldstein Ferber, *Very Early Clocks*,
https://doi.org/10.1007/978-3-031-89603-3_7

Seminal studies on early perception of music and verbal stimuli transmitted as audio signals from the out-uterine environment or the abdomen of the pregnant mother to the fetus showed that the later born infant reacts to these stimuli with behavioral reactions interpreted as familiar by measures of heart rate and heart rate variability (Shuffrey et al., 2019).

These studies and others led the scientific community to a large agreement that beyond early fetal behavior, the fetus is able to perceive early sensory stimuli transmitted from outside the womb around the time of cortical somatosensory area early maturation. Most of these studies were carried out after mid-gestation; thus, these fundamental abilities of the fetus are regarded as the building blocks of the developing qualities of transitory awareness, gradually preparing the fetus to the out-uterine environment.

Since these data show that newborns were able to detect and respond to stimuli following exposure to these stimuli as fetuses, several questions that arise here are: What is an early perception and how can it be validated? What is the gradual clockwise molecular process of the earliest perceptions as they become more and more complex and related to early behavior? How do fetal executed behaviors shape synaptogenesis? How do the rhythms of the single cell coordinate with the rhythms of wiring with other neurons? This chapter will address these questions.

Perception, even if defined as an early quality of awareness (Herlenius & Lagercrantz, 2004), requires the functions of neural connectivity and neural networks. The relevant, partially overlapping in time sequences include cell differentiation, migration to the subcortical plate area, cell specification, and cellular competition towards the onset of pruning which is experience-shaped. Early sensations, when repeated, are followed by fetal movements, which in turn support expanding of the movements and repeating them. This initiates, when complex enough, shaping of synaptogenesis in the experienced direction.

To support the claim that this type of experience-shaped synaptogenesis involves early perceptions, it is noted that recent data show the phenomenon of fetal habituation to repeated stimuli and fetal orienting response to an intrauterine stimulus, in addition to older studies showing these capacities in early-born premature infants before reaching term age. It has been accepted in science that these phenomena require short-term memory representations (Donovan et al., 2020; Herlenius & Lagercrantz, 2004).

Recent studies on early behavior show that the embryo exhibits random stereotyped behaviors currently understood by scientists as exploration and novelty seeking. These behaviors increase the future fetus's knowledge on its own body and its interface with the intrauterine environment in terms of time, space, and general body surface structure (Fagard et al., 2018). From these general early motor behaviors, termed by others as "motor bubbling," arise, around mid-gestation, other complex intentional behaviors such as self-touching and especially limb touch of highly already innervated areas, such as the face. This is also the timing of coupling between heart rate, sleep state discrimination, and motor behavior (Ferber et al., 2023), showing a more integrated reaction of the early CNS than during embryonic periods.

It is widely accepted that the quality of neural wiring is crucial for any human perception to occur. Interestingly, the rise of inhibitory factors in fetal development at mid-gestation marks the first network wiring and its more organized behavioral outcomes. Recent, as well as older, studies show that a coherent function of a neural network depends on the neural cell's stable final location and specification as well as on cellular competition to provide neural signaling as "firing together." This complex neural functionality develops through an order set upon the random embryonic signaling by "senior" older neurons providing a "stop signal" to new "younger" neurons (Fischi-Gómez et al., 2015; Jaimes et al., 2020; Goldstein Ferber et al., 2022; Tau & Peterson, 2010; Dehorter et al., 2012). The "stop signal" is available to older neurons at the timing of rise of inhibitory factors during gestation.

The importance of the "stop signal," or inhibition, in fetal development requires understanding of its molecular origins. Notably, neural polarization and depolarization precedes the development and function of neurotransmitters. Additionally, inhibitory factors such as NMDA mediated by AMPA turn to be crucial for healthy fetal development from mid-gestation onwards (Herlenius & Lagercrantz, 2004). Alterations in NMDA channels, including their impact on Mg^+ blockade, serve to protect against risks of excitotoxicity, at a time period in gestation when the overexcitation is not proper as before. NMDA functions which show differences in "older" vs. "younger" neurons present NMDA alterations as time-sensitive (Pfisterer & Khodosevich, 2017; Rahman et al., 2017).

The occurrence of the GABA function in the embryo shapes synaptogenesis. Levels of GABA in the embryo are modulated by intracellular chloride, while GABA is considered the first contributor to neural network formation. GABA and its excitatory function in the embryo, elicited by a complex ionic function including the cotransporters of Cl^-, KCC1, and NKCC2, balance cellular Ca^{2+} efflux and protect the osmotic homeostasis of the cell. Younger neurons remove Cl^- less efficiently than older ones. The excitatory function of GABA in the embryo is dependent on the dynamics of intracellular Cl^- cotransporters and their timely impact on polarization. Later, between 25 and 30 weeks of gestation, the function of excitatory glutamate appears together with a reduction in the earlier excitatory function of acetylcholine (Goldstein Ferber et al., 2022). This later occurrence is due to its dependence on full morphological development of the neuron, including basal and apical dendrites and axonal growth up to a rate that enables them to reach upper molecular layers to balance the osmo-cellular equilibrium, without interfering with the originally preprogrammed white matter growth. Thus, the excitatory GABA-glutamate interplay starts in fully developed neurons, and it switches to determine inhibition-excitation balance through Cl^- mobilization of calcium ions only around mid-gestation. This shows the crucial role of chloride in mobilizing the neural cell towards readiness for depolarization and coordinated firing for wiring together in networks though postsynaptic coherent transmissions.

The first immature coordinated currents were observed in the immature patterns of firing together elicited as voltage-gated high frequencies of calcium efflux in GABAergic interneurons. This advance is necessary for generation of genetic cascades and stopped by blockers of gap junctions after reaching calcium plateaus.

Giant developing potentials are considered the first synapse-driven neural networks. Hub neurons were observed to discharge before GABAergic and glutamatergic neurons, during giant polarization establishment (Dehorter et al., 2012).

Interestingly, recent data suggest that the rhythms of cellular osmosis are controlled by clock genes through their rhythmic posttranslational regulation of protein activity, ion transport, and temporal consolidation of proteome renewal (Stangherlin et al., 2021). The changes in the Cl^- dynamics and clock genes' control of the cellular osmosis at mid-gestation coincide with the timing of synaptic changes driven by fetal activity and with the timing of coordinated firing of networks. This is also the timing at which more organized and complex behaviors occur in gestation. The timing of these associations may imply the onset of early interplay between excitation and inhibition following embryonic stages of excitation dominance. This may be a prerequisite for the elicitation of partial awareness incidents. This is in accordance with recent data showing that any "freeze" in these timewise sequences, including any delays in the occurrence of its peaks, results in misconnections between "older" and "younger" neurons with subsequent consequences of neurological postnatal disorders and insults to white matter (Fischi-Gómez et al., 2015; Goldstein Ferber et al., 2022; Jaimes et al., 2020; Tau & Peterson, 2010).

According to recent findings, the outgrowth of thalamic fibers toward the somatosensory area, one of the structural healthy brain development hallmarks, precedes in 2 weeks the outgrowth of other thalamic fibers to other developing areas (Kostović et al., 2021). This implies that the neural circuitry responsible for the experience of sensory stimulation such as tactile, auditory, temperature, and pressure precedes generation of early perceptions which occurs later between 24 and 30 weeks of gestation, with the penetration of thalamic fibers into the cortical plate. In this time window, the onset of fetal body exploration in the form of self-touching has been observed by 3D ultrasound (Einspieler et al., 2021; Fagard et al., 2018), suggesting also the start of the "body-map" creation as short-term memory chunks (Fagard et al., 2018; Lagercrantz & Padilla, 2016). Other reports show that self-touching behaviors at this timing in gestation stimulate the growth of synapses and contribute to the shaping of synaptogenesis (Sciences et al., 1992). From this point in gestation and onwards, the fate of the neurons firing together and wiring together, including cell specification and their determination, depends also on their previously produced behavioral outcome. Early sensations may then be shaped into early perceptions if the complex of wiring matches a repeated behavioral outcome.

This behavior-wiring feedback loops result from shifts from original, embryonic, molecular, and chemical functions which were determining and shaping the future neuron's fate, including the rise of inhibitory factors at earlier stages of gestation. The shift at mid-gestation requires a larger repertoire in "older" neural networks than in "younger" network to establish a neural network. Fetal behaviors shape this developing repertoire, and earlier born neurons serve as a "stop signal" to later born neurons firing (Dehorter et al., 2012).

The generation of neural networks within and between the developing regions precedes the horizontal cortico-cortical coordinated firing. Differences in speed of maturation were found between cortical deep layer neurons which were generated first, before 16 weeks of gestation, compared to the maturation of upper layer neurons that followed and accumulated up to 21 weeks of gestation (Fan et al., 2020). Thus, excitatory neurons in the developing cortex across cortical lobes develop in a non-coordinated but time-sensitive manner.

These differences in the speed of maturation may be attributed to the timing of transcription factor expression. For example, the transcription factor CTIP2$^+$ was found to be dominant in the cortical plate at 13 weeks of gestation, while the transcription factor SATB2$^+$ was dominant at 20 weeks of gestation at the same developing brain location. However, a reduced number of SATB2$^+$ were found during deep layer neuron generation in the frontal lobe compared to more such neurons in the temporal lobe, suggesting different molecular regulation of neurogenesis in each developing cortical region. Higher neural enrichment in the frontal lobe by 24–25 weeks of gestation, especially by motor neurons, shows more rapid maturational development in the frontal lobe compared to other lobes. These data indicate specification of regional development in terms of speed of wiring for each cortical area, separately. Accordingly, the nuclear receptors NR2F1 and NR2F2 were found earlier than 20 weeks of gestation to be by far more expressed in the temporal lobe during neural progenitor cell differentiation into excitatory neurons, while larger enrichment of deep layer neurons was observed at 20 weeks of gestation in the frontal lobe (Fan et al., 2020). These findings suggest that molecular balances appear at the single cell level in parallel with regional development and that these asynchronous but coordinated maturational processes are the complex underlying the elicitation of wiring within the young cortex, thus defined as the underlying molecular basis of the early transient quality of awareness and the onset of perceptual clocks.

These findings directly match findings on firing rate and wiring levels at 24 weeks of gestation as well as findings on the efficiency of the Na^{2+} efflux at this age, with a significantly larger amplitude than earlier in gestation. Also, spontaneous excitatory postsynaptic currents were found to be more frequent in the frontal lobe compared to those found in the other lobes, suggesting that wiring occurs in the frontal lobe earlier than other developing cortical regions and reaches a higher level of network maturation by 24 weeks of gestation. Thus, the accuracy of molecular factors of the single cell facilitates periods of the frontal lobe's rapid expansion. A striking example are the newly identified transient neural progenitor cells (Fan et al., 2020), with quick proliferation and differential capacities first recorded between 11 and 14 weeks of gestation, while showing a marked developmental impact on deep and upper layer neurons from mid-gestation onwards.

In addition to short- and long-distance simultaneous neural firing, recent studies show that macroglia and microglia cells contribute to neural network wiring and speed of transmission within regions in the cortical plate and between rostral and caudal areas. Glial cells lie in close proximity to neurons, precursors, and neural progenitors, and interact with all these types of cells as well as between themselves. These cells are able to define atypical potential errors in wiring and cell survival earlier than the maturation of any specific wiring. They also regulate the osmotic homeostasis in the neural cell through choroid plexus macrophages and promote myelination; As such, they contribute to the speed of wiring. Interestingly, recent findings show that the choroid plexus, a small monolayered epithelium located in the brain ventricles that serves to secrete the cerebrospinal fluid, contributes to early apical development in rhythms determined by the oscillations of clock genes (Quintela et al., 2021). This cascade, originating in the clock gene ultradian oscillations, is afforded by their coordination with the choroid control of ion channels, especially that of the potassium ion intra- and extracellular concentrations, and transepithelial major electrolytes, Na^+ and Cl^- (MacAulay & Toft-Bertelsen, 2023).

Glial cells show a very early origin starting with the creation of the yolk sac and early stages of the premature CNS during the embryonic stage as shown in animal and human studies, but their development spans the gestational period and even continues through the postnatal early phases. Their development occurs through feedback loops with neural cells that in turn develop through the glia impact. Glial cell signaling is essential for neuronal pathfinding during migration, while neurons' local reactive signaling regulates colonization patterns in both time and space as well as heterogeneity of glial cell identity during their differentiation (Thion et al., 2018).

Along their developmental trajectory, neural progenitors continue to produce oligodendrocytes in the parenchyma while the microglia undergo local differentiation. Maternal inflammation changes their developmental trajectory and their preprogrammed pathways, thus posing a risk to the development of wiring and the resulting clocks of perception. Their most significant role for wiring networks within and between the premature brain areas is in guiding migration and neural pathfinding. Therefore, glia's role before the closure of the blood-brain barrier (day 13 of gestation in rats) is crucial for the successful penetration of the thalamic fibers into the cortical plate at 24 weeks of gestation in humans (Kostović et al., 2021).

Glial cells were recently identified also as responsible for molecular cell maturation, and as such they contribute to both parallel sequences of the single cell coinciding that of the regional and interregional wiring development. A recent comprehensive study of humans showed a developmental trajectory of oligodendrocyte lineage cells from embryonic stages of 7 weeks of gestation until mid-gestation and term (38 weeks post-conceptual age) (Fan et al., 2020). By comparing cortical regions to the pons at 20 weeks of gestation, much higher abundance of astrocytes has been found in the pons than in the developing cortex, while complex processes were captured in the cells located in the pons. This suggests earlier and quicker maturation of this type of cells in lower regions with suspected differences in speed of maturation as well as the effects of bottom-up signaling and migration.

However, classical (Leone et al., 2008) and recent studies (Fan et al., 2020) also showed that lower regions, such as the pons, contain less projection neurons than higher regions such as the cortical plate, at least between the end of the embryonic stage and mid-gestation. These data suggest that projection neurons have a special role in wiring within and between higher and lower regions by projecting from upper regions down to meet migrating cells from lower regions along their pathfinding upward, towards further maturational stages at 24–25 weeks of gestation. The opposite has been found for inhibitory neurons with higher abundance in the lower regions earlier than found within the cortical plate, suggesting local cortical origination of projection neurons contributing to neural wiring. This also suggests an expanded preprogramming time for rostral to caudal migration of inhibitory neurons, starting at early mid-gestation. Although there are marked differences in the molecular single-cell developmental timeframes as well as in the cortical regions' differential speed of maturational parallel sequences, these complex processes reach a peak of wiring to support early perceptions around mid-gestation at 24–25 weeks post-conceptual age.

The conclusions on the trajectory and underlying mechanisms of early perception clocks carry implications for future diagnostic tools of screening growing fetuses, especially towards the gestational timing of reaching the peak of fetal heart-rate-sleep-states-behavior coupling, which I termed earlier in an evidence-based manner: "when the mind comes to live inside the body" (Ferber et al., 2023). It is concluded here that early perceptions depend on parallel sequences of single-cell, regional, and interregional development, in an asynchronous manner although sharing the same time period. These currents gradually become more coordinated than during embryonic phases.

Thus, early perceptions are not a simple result of cortical white matter growth, nor do they depend on one particular structural or functional component or only on sensations and the somatosensory regional maturation. They are also not the direct result of firing together or wiring together of neurons. Rather, the clocks of perception depend on the timing of molecular, regional, and interregional factorial convergence as it occurs in concert and in relation to a given behavior at a given preprogrammed time along gestation. As such, the clocks of perception heavily rely on progenitors and cotransporters in rhythms controlled by clock genes for osmotic regulation of the cell during early gestation.

Parallel sequences are added from late embryogenesis to mid-gestation, consisting of the rise in inhibitory factors and neuronal migration guided by glial cells in pathfinding at specific timings. This is accompanied by the onset of circadian rhythms in the life of each tissue and by the shift in clock gene oscillations from cycling in ultradian rhythms only. This process includes neuron-glia feedforward loops resulting in region-specific rhythms of structural growth.

In healthy gestation, this multifaceted complex produces incidences in which upon a given timing of neurons firing together, they wire together in relation to a given fetal behavior at a specific timing of intrauterine conditions. From mid-gestation onwards, the number of these incidents gradually increases along the time flow towards term. With such a time-sensitive infrastructure, the clocks of perception follow a gradual widening pathway depicted by active collection of germinal transitory quality of awareness peaks, resulting from sustained wiring, expanded neural repertoire through synapse selection, for a given behavior at a given compound of intrauterine conditions from mid-gestation onwards. These peaks in turn lead to an expanded range and complexity of more organized and intentional behaviors, characterized by heart-rate-sleep-states-behavior coupling, on the expense of behavioral frequencies and random stereotyped movements, which belong to earlier stages of gestation and lack these aligned quality of transitory awareness. In this sense, very early perceptions occur in very short instances when molecular events meet morphological readiness to exhibit a behavior and absorb its results at the same time. The expansion of these instances depends on the rhythms of their repetition.

References

Dehorter, N., Vinay, L., Hammond, C., & Ben-Ari, Y. (2012). Timing of developmental sequences in different brain structures: Physiological and pathological implications. *European Journal of Neuroscience, 35*(12), 1846–1856. https://doi.org/10.1111/j.1460-9568.2012.08152.x

Donovan, T., Dunn, K., Penman, A., Young, R. J., & Reid, V. M. (2020). Fetal eye movements in response to a visual stimulus. *Brain and Behavior, 10*(8), e01676. https://doi.org/10.1002/brb3.1676

Einspieler, C., Prayer, D., & Marschik, P. B. (2021). Fetal movements: The origin of human behaviour. *Developmental Medicine & Child Neurology, 63*(10), 1142–1148. https://doi.org/10.1111/dmcn.14918

Fagard, J., Esseily, R., Jacquey, L., O'Regan, K., & Somogyi, E. (2018). Fetal origin of sensorimotor behavior. *Frontiers in Neurorobotics, 12*. https://doi.org/10.3389/fnbot.2018.00023

Fan, X., Fu, Y., Zhou, X., Sun, L., Yang, M., Wang, M., Chen, R., Wu, Q., Yong, J., Dong, J., Wen, L., Qiao, J., Wang, X., & Tang, F. (2020). Single-cell transcriptome analysis reveals cell lineage specification in temporal-spatial patterns in human cortical development. *Science Advances, 6*(34), eaaz2978. https://doi.org/10.1126/sciadv.aaz2978

Ferber, S. G., Geva, R., & Weller, A. (2023). When the mind comes to live inside the body: The ontogeny of the perceptual control clock. *Current Neuropharmacology, 21*(1), 13–21. https://doi.org/10.2174/1570159X20666220411095508

Fischi-Gómez, E., Vasung, L., Meskaldji, D.-E., Lazeyras, F., Borradori-Tolsa, C., Hagmann, P., Barisnikov, K., Thiran, J.-P., & Hüppi, P. S. (2015). Structural brain connectivity in school-age preterm infants provides evidence for impaired networks relevant for higher order cognitive skills and social cognition. *Cerebral Cortex, 25*(9), 2793–2805. https://doi.org/10.1093/cercor/bhu073

Goldstein Ferber, S., Weller, A., Ben-Shachar, M., Klinger, G., & Geva, R. (2022). Development of the ontogenetic self-regulation clock. *International Journal of Molecular Sciences, 23*(2), Article 2. https://doi.org/10.3390/ijms23020993

Herlenius, E., & Lagercrantz, H. (2004). Development of neurotransmitter systems during critical periods. *Experimental Neurology, 190*, 8–21. https://doi.org/10.1016/j.expneurol.2004.03.027

Jaimes, C., Machado-Rivas, F., Afacan, O., Khan, S., Marami, B., Ortinau, C. M., Rollins, C. K., Velasco-Annis, C., Warfield, S. K., & Gholipour, A. (2020). In vivo characterization of emerging white matter microstructure in the fetal brain in the third trimester. *Human Brain Mapping, 41*(12), 3177–3185. https://doi.org/10.1002/hbm.25006

Kostović, I., Radoš, M., Kostović-Srzentić, M., & Krsnik, Ž. (2021). Fundamentals of the development of connectivity in the human fetal brain in late gestation: From 24 weeks gestational age to term. *Journal of Neuropathology & Experimental Neurology, 80*(5), 393–414. https://doi.org/10.1093/jnen/nlab024

Lagercrantz, H., & Padilla, N. (2016). The emergence of consciousness: From foetal to newborn life. In *Neurotechnology and Direct Brain Communication*. Routledge.

Leone, D. P., Srinivasan, K., Chen, B., Alcamo, E., & McConnell, S. K. (2008). The determination of projection neuron identity in the developing cerebral cortex. *Current Opinion in Neurobiology, 18*(1), 28–35. https://doi.org/10.1016/j.conb.2008.05.006

MacAulay, N., & Toft-Bertelsen, T. L. (2023). Dual function of the choroid plexus: Cerebrospinal fluid production and control of brain ion homeostasis. *Cell Calcium, 116*, 102797. https://doi.org/10.1016/j.ceca.2023.102797

Pfisterer, U., & Khodosevich, K. (2017). Neuronal survival in the brain: Neuron type-specific mechanisms. *Cell Death & Disease, 8*(3), e2643–e2643. https://doi.org/10.1038/cddis.2017.64

Quintela, T., Furtado, A., Duarte, A. C., Gonçalves, I., Myung, J., & Santos, C. R. A. (2021). The role of circadian rhythm in choroid plexus functions. *Progress in Neurobiology, 205*, 102129. https://doi.org/10.1016/j.pneurobio.2021.102129

Rahman, T., Zavitsanou, K., Purves-Tyson, T., Harms, L. R., Meehan, C., Schall, U., Todd, J., Hodgson, D. M., Michie, P. T., & Weickert, C. S. (2017). Effects of immune activation during early or late gestation on N-methyl-d-aspartate receptor measures in adult rat offspring. *Frontiers in Psychiatry, 8*. https://doi.org/10.3389/fpsyt.2017.00077

Sciences, N. A. of, Medicine, I. of, & Ackerman, S. (1992). *Discovering the brain*. National Academies Press.

Shuffrey, L. C., Myers, M. M., Odendaal, H. J., Elliott, A. J., du Plessis, C., Groenewald, C., Burd, L., Angal, J., Nugent, J. D., Isler, J. R., & Fifer, W. P. (2019). Fetal heart rate, heart rate variability, and heart rate/movement coupling in the safe passage study. *Journal of Perinatology, 39*(5), 608–618. https://doi.org/10.1038/s41372-019-0342-9

Stangherlin, A., Seinkmane, E., & O'Neill, J. S. (2021). Understanding circadian regulation of mammalian cell function, protein homeostasis, and metabolism. *Current Opinion in Systems Biology, 28*, 100391. https://doi.org/10.1016/j.coisb.2021.100391

Tau, G. Z., & Peterson, B. S. (2010). Normal development of brain circuits. *Neuropsychopharmacology, 35*(1), 147–168. https://doi.org/10.1038/npp.2009.115

Thion, M. S., Ginhoux, F., & Garel, S. (2018). Microglia and early brain development: An intimate journey. *Science, 362*(6411), 185–189. https://doi.org/10.1126/science.aat0474

Chapter 8
The Clocks of Motivation: Rhythms of Molecular Restraints Originate Intentional Behaviors

Abstract Contrary to the thought that early motivation is based on an increasing drive to grow and move forward, this chapter integrates findings showing that accumulation of more organized, intentional, and stimulus-related fetal behaviors occurs from mid-gestation onwards based on multileveled inhibitory factors. As such, the very early clocks of motivation are driven and timed by accurate elicitation of complex inhibitory factors. These include transcriptional and molecular factors and their behavioral outcomes. Secreted antagonists of bone morphometric proteins were found to bind to it in the extracellular space and inhibit activation of its receptor resulting in the first formation of the neural tissue at 18 weeks of gestation. COX-1 and COX-2 enzymes protect against apoptosis, and Mg^{2+} blockade becomes a balancing factor of neural firing rate reducing the earlier embryonic overshooting. The GABA switch from excitatory to inhibitory affords later waves of migration. Later, radial glia is eliminated to afford the complete formation of thalamic fibers and their invasion of the cortical plate, including their profound impact on deep associative neurons at this time of invasion. Later in gestation, the EEG alpha wave representing awake and partial awareness states occurs on the expense of delta waves detected during earlier stages of gestation. On the behavioral level, more complex and intentional fetal reactions occur through inhibition of earlier types of behaviors in favor of new phenotypes as shown in novel 3D and 4D ultrasonic techniques. Clock genes govern the rhythms of inhibitory factors though their heterodimerizing effects and modulation of the excitatory Ca^{2+} entries to the cell. The capacity of clock genes to converge unstable molecules to stable ones reduces the amounts of oxidative stress and elicits antioxidative defenses, thus promoting the rise of excitation-inhibition balance during the transition of mid-gestation. By their noted capacities, clock genes protect the mitochondrial functions to be appropriate for the gestational age metabolic demands; thus, their functioning rhythms protect the advance in structural growth. This in turn supports more efficient fetal recovery from unpleasant intrauterine conditions and inhibition of these conditions' aligned behaviors. It is concluded that at mid-gestation, when the onset of excitation-inhibition rhythms meets structural accomplishments appropriate for this gestational age, through a complex of inhibitory factors, elicitation of intentional behaviors occurs on the expense of earlier more frequent random and stereotyped behaviors.

S. Goldstein Ferber, *Very Early Clocks*,
https://doi.org/10.1007/978-3-031-89603-3_8

Motivation is theoretically depicted as driven by frustration or threat on the one hand and by the drive to move forward to further steps of development on the other hand (Wasserman & Wasserman, 2020). If so, how is early motivation experienced? What is experienced as frustration and threat by the fetus? What moves the fetus to further behavioral development? What are the generators and controllers of early clocks of motivation?

This chapter will show the contribution of inhibition on both the molecular and the behavioral levels for healthy development of early motivational clocks reacting to both the frustration and the drive to move forward in the embryo and fetus. In this sense, frustration is defined as a stop signal for the drive to move forward, and as such it retains vital and evolutionally benefits to support healthy early human functioning by eliciting excitation-inhibition balance. From the frustration angle, the womb does not fully protect the embryo and fetus, and unpleasant conditions may occur in the form of oxidative stress with compensatory reversing enzymatic mechanisms for continuation of growth and preservation of the move-forward excitatory progress.

Extreme frustration such as early pain reactivity has been demonstrated and validated in fetal (Lagercrantz & Changeux, 2009) and preterm infant life (Grunau, 2020), following the borderline of viability at 24 weeks of gestation (Kostović et al., 2021). At this gestation age, intentional and stimulus-responsive fetal behaviors have been recorded (Donovan et al., 2020). This is the timing of very early sensorimotor integration, although this integration is not necessarily representing the onset of sensing and reacting to unpleasant intrauterine conditions nor the start of inhibitory factors on the molecular and behavioral levels. That is, a fetal intentional behavior at mid-gestation is built upon a long molecular pathway of changes and shifts, responding to earlier rhythms of development and growth.

The move-forward drive starts at early embryonic life, and it is affected by sympathetic activity from early stages of gestation through the first year of life. Sympathetic dominance is apparent at the start of this system's development, emerging from the early developing brainstem and before the emergence of the inhibitory parasympathetic system from the same organ (Goldstein Ferber et al., 2022). This dynamic process advances from proliferation and differentiation to the drive for the first sucking behavior, first autonomous breathing, motor development through the first step, and the first word at around 1 year of life. That is, early motivation represents the drive to grow and develop complex neural and behavioral systems.

It is accepted that cortical sensorimotor development, starting at 18 weeks of gestation and maturing at 28 weeks of gestation (Afif et al., 2015), is the neurodevelopmental process responsible for the pathways by which motivation arises. This implies that sensors and behavioral activators, which when combined represent the axis of early motivation, are apparent as early as embryonic life and peak through mid-gestation. The clocks of this early motivational axis develop concomitantly with the clocks of the neural tube, from the first neural cell up to the stage of neural tube closure. The ability of the healthy fetus to feel pain upon maturation of these two concomitant developmental processes at 28 weeks of gestation is an accomplishment based on the progressing development of the gradually specified cellular sensors and activators, termed by others "intrinsic motivation" (Fagard et al., 2018). This does not require integration of sensing nor awareness up to mid-gestation, and therefore the term "experience" belongs to the second half of human gestation.

Clock genes govern the rhythms of inhibitory factors though their heterodimerizing effects and modulation of the excitatory Ca^{2+} entries to the cell. The capacity of clock genes to converge unstable molecules to stable ones reduces the amounts of oxidative stress and elicits antioxidative defenses, thus promoting the rise of excitation-inhibition balance during the transition of mid-gestation. This in turn supports more efficient fetal recovery from unpleasant intrauterine conditions and inhibition of their associated behaviors. The rise of circadian rhythms in every fetal tissue marks the onset of "experience" upon the borderline of viability. Earlier ultradian oscillations of clock genes is designed for the on-off switches of arousal and decline (van der Veen & Gerkema, 2017) and serves as the molecular building blocks of early motivation clocks.

I argue that (1) early motivation is molecularly based; (2) the clocks of motivation are apparent in embryonic and fetal motor behavior and its alterations along gestation, within a progress of eliminating earlier patterns in favor of new ones (Einspieler et al., 2021; Fagard et al., 2018); (3) the rise of inhibition during gestation is apparent in – and contributes to early motivation on both the molecular and behavioral levels; and (4) the rhythms of motivation clocks development are governed by clock gene early and late oscillations.

The first question that arises at this point is: how is early motivation molecularly based? Neuronal polarization during embryonic life may be viewed as the drive to move forward with rapid high-frequency progenitors signaling from conception onwards. To this, high frequency of neural firing rate is added from the start of neural tube construction at 18 weeks of gestation. This is apparent through Ca^{2+} efflux. The overexcitation by this efflux causes the rise of Mg^{2+} blockade in the neural cell resulting from signaling loops with the related progenitors. This point is among the topics yet to be studied. It was mentioned in a recent publication that the Mg^{2+} blockade rises from the "shade" of Ca^{2+} efflux (Komiya & Runnels, 2017). Thus, through inhibition by the Mg^{2+} blockade and the excitatory impact of the Ca^{2+} flux entries to the cell, neural polarization-depolarization periods are established. Thus, Mg^{+} coordinated by clock gene early oscillations is among the first inhibitory agents.

From another angle, and in order to show the contribution of molecular inhibition, it is noted that the excessive firing incidents during embryonic life are associated with random stereotyped behavioral movements during the first 2 weeks of post-conceptual age (Fagard et al., 2018). This shows the significant contribution of Mg^{2+} blockade to the excessive firing rate as a balancing factor of overshooting, which in turn is responsible for random behaviors during embryonic stages of development. Others showed that by expression of the Mg^{2+} transporters TRPM6 and TRPM7 in developing embryos, Mg^{2+} plays a critical role in gastrulation and neural fold closure even later than 28 weeks of gestation, and its deletion is followed, at least in mice, by lethality of neurons and neural tube defects up to complete absence of neural tube closure (craniorachischisis) (Komiya & Runnels, 2017).

Furthermore, as early as the 1990s of the previous century, secreted antagonists of bone morphometric proteins (BMPs) such as Noggin, Chordin, Follistatin, Cerberus, and XNr3 were discovered. They were found to bind to BMP in the extracellular space and inhibit activation of its receptor. As a reaction, the naïve embryonic ectoderm forms the neural tissue and not epidermis at 18 weeks of gestation. In conditions of blocking these inhibitors, the neural cell did not form. To show the significance of these inhibitors for healthy development of behavior, it has been shown that the anterior neural genes are directly induced when these inhibitory proteins are activated (Rogers et al., 2009), showing again, based on these additional data, that inhibitory factors are heavily involved in the healthy generation of the neural tube and its closure.

Thus, the governing of behavioral motivation in early intrauterine life implies that extra excitation and firing rates precede inhibitory functions and the excitation-inhibition balance, which occur only from mid-gestation onwards. This means that extra excitation aimed at structural growth, including proliferation, is needed in early embryonic life, whereas inhibitory factors are needed for healthy development of functionality including cell differentiation, migration, and specification later in gestation. With regard to motivation, early clocks of motivation are based on heightened arousal of the primitive organism, but its behavioral complex applications appear to be due to inhibitory factors as they become available later in fetal life. This molecular basis shows the significant role of the neural "stop signal" termed in behavioral sciences "frustration" when describing postnatal and adult behaviors.

Inhibition of behavioral patterns arises also from the molecular basis, involving changing amounts of oxidative stress due to mitochondrial synthesis for free energy (Ferber et al., 2024) and in more severe cases due to maternal stress or short-term insufficiencies of the placenta as synthesized in the labyrinth zone. Through protective functions of enzymes such as COX-1 and COX-2, to avoid apoptosis (Burdon et al., 2007), healthy behavioral patterns and their developmental alterations may progress. The clock gene rhythms guide such enzymatic and nonenzymatic antioxidative defenses which show a crucial impact of inhibiting exhaustion of the mitochondria, while keeping its molecular age-dependent demands intact from mid-gestation onwards.

The short-term manner and the quick recovery from fetal unpleasant conditions may retain the positive role of inhibition shown in elimination of early behavioral patterns in favor of novel patterns and the consolidation of depolarization rates as well as synapse selection, pruning, and cell specifications. That is, the womb does not fully protect the fetus, but even as such, it has productive contributions. If the fetus experiences a range of unpleasant conditions, that may be understood as early "frustration," not reaching extremes and modulated molecularly, forming the available move-forward organismal motivational systems, gestational age-dependent.

The role of inhibition in the development of embryonic and fetal behavior has been discovered in recent years with the novel technology of 3D and 4D ultrasound.

It was shown that the pattern of fetal behavioral movements and reactions changes along gestation. An earlier pattern is inhibited for a novel pattern to emerge (Fagard et al., 2018). The embryo starts from random stereotyped movements up to repeated complex fetal movements which increase in frequency as the birth becomes closer in time. These changes eliminate the earliest pattern when a new pattern is developed, showing the contribution of inhibition for further behavioral growth. Heart-rate-sleep-states-movements coupling has been observed around mid-gestation, and complex repeated behaviors, even intentional, have been recorded with elevated frequency as gestation progresses towards birth (Fagard et al., 2018; Ferber et al., 2023).

Note that, in total, frequencies of fetal behaviors are reduced compared to embryonic movements but are more complex, and some of them are even intentionally repeated. Spontaneous firing of motor neural cells begins its signaling at the same time they differentiate (Wilkinson et al., 2017). In fact, motor activity starts during embryonic life as a result of rhythmic bursts of thousands of cells' generated action potentials searching for their synapse selections and signaling networks while innervating skeletal muscles. How does this process happen on the neurobehavioral level? What type of behaviors is inhibited and when?

The first movements were observed as early as 7 weeks of gestation. These first movements are side bending of the head or of the rump, stereotyped movements which soon disappear and give rise to new ones. Startles followed by general movements increase their frequency from 8 weeks of gestation onwards and decrease at 17 weeks of gestation. After the 17th week of gestation, general movements do not occur compulsively following startles. Isolated movements emerge after the occurrence of general movements at 14 weeks of gestation. Twitches during sleep occur at 10–12 weeks of gestation increasing from 15–16 weeks of gestation until sleep states consolidate at mid-gestation, and periods of quiescence extend from 20 weeks of gestation onwards (Fagard et al., 2018; Ferber et al., 2023; Goldstein Ferber et al., 2022). At these stages, movements do not give the impression of aiming at a precise goal nor for eliciting sensation. This occurs later on at 16–18 weeks of gestation.

At 20–24 weeks of gestation, touching hand to hand and crossing hands is observable, eliminating previous reflexive behavior. Self-touching of the eye and the mouth with increased repetitions of slower movement dynamics of arm to the eye, compared to quicker dynamics of movement of arm to the mouth, is observable from mid-gestation onwards. This is an interesting finding of showing that the fetus is able to selectively touch the sensitive region of the eyes compared to touching the mouth region. Touching of the mouth region includes the pleasure of sucking the thumb, as preparation for out-uterine feeding, a pleasure that the fetus seeks to quickly exercise and repeat. Both types of repeated touching are aimed at heavily innervated regions by that time of gestation, the eyes and the mouth (Fagard et al., 2018).

The most frequent movements from mid-gestation onwards, which concomitantly meet the timing of the neural tube closure and integration of the sensorimotor regions, are facial expressions. These complex repeated movements eliminate movements that were apparent in the previous stages and increase their complexity around the period of 25–34 weeks of gestation, when the cortical plate is organized in six layers and axons from other regions reach it.

During late periods of gestation, early motivation for sensation-oriented goals is repeatedly shown, termed by others "the exploration of a body map" (Fagard et al., 2018). All the other preceding stages that are developmentally inhibited by mid-gestation showing that the clocks of early motivation emerge from random firing but develop to a complexity of behaviors shaped by evolution to meet the goals of preparation towards social communication as the birth timing comes closer.

The early motivation for sensation-oriented behavioral incidents coincides with the neural tube closure timing. How is this accomplishment achieved, as it starts its development from random behavioral movements? How may this accomplishment be explained as a process rather than just a sudden change on the neural structural and functional levels?

While transcription factors guide and lead the feedforward excitatory signaling to the differentiating and migrating cells during earlier stages, the consolidation of inhibition on the molecular and connectivity pathways may explain how the seeds of early motivation grow along early stages of gestation. Indeed, the process of neuronal migration to the cortical plate starts with glutamatergic excitatory projection neurons, while one of the major migratory indirect events occurs in the ganglionic eminence routes that contain GABA interneurons after the switch of GABA at mid-gestation from being excitatory to its traditional known role as inhibitory (Goldstein Ferber et al., 2022; Wilkinson et al., 2017). These findings show the dominance of excitation and the move-forward drive during early embryonic developmental stages and the gradually mounting inhibition along gestation, which affords later waves of migration following the GABA switch to inhibitory, especially following the mid-gestation timing (Goldstein Ferber et al., 2022).

Interestingly, the timing of the main migratory events towards the thalamus shows that neuronal migration onset occurs in humans on the route from the ventricular zone toward the pia mater around 33 days of gestation (Bystron et al., 2008), while the major migratory activity has been shown between 12 and 20 weeks of gestation, with a distinct movement from the ganglionic eminence towards the dorsal thalamus, finally with a definable formation by 25 weeks of gestation. By 26 weeks of gestation, radial migration is nearly complete, and radial glia are replaced by astrocytes and thalamic afferent fibers which start to develop (Xu et al., 2014). By 36 weeks of gestation, close to the timing of expected birth, radial glia disappears and afferent thalamic fibers are fully formed (McKinstry et al., 2002). The disappearance of radial glia is under the control of clock genes when they oscillate in circadian rhythms at the ending stage of neurogenesis (Arafa & Emara, 2020). This is an additional inhibitory factor which eliminates the radial glia, by the "off" option in the "on-off" capacity of clock genes, for the thalamic fibers to move forward, complete their form, and invade the cortical plate. This accomplishment shows that the crucial neuronal migratory events become more significant as the time of gestation flows and that the inhibitory factors combine with excitatory ones later in gestation. Although migration is an excitatory function, its major portion grows and becomes significant by accumulation of inhibitory factors.

It is well accepted that the maturational process of the sensorimotor region is the process responsible for sensation-oriented goals of fetal behavioral patterns. The first stage of this process occurs at 18 weeks of gestation when the inferior part of the central cerebral sulcus appears. The second stage occurs at 20 weeks of gestation with the development of the lateral regions. The third stage occurs at 24 weeks with the development of parietal and temporal cortices covering the insular region. The ending phase of this maturational process occurs between 26 and 34 weeks of gestation with the final phase of thalamic fibers' penetration into the cortical plate including the completion of migration and formation of the thalamic fibers which are processes coinciding with the resolution of the subcortical preparatory plate (Kostović et al., 2021; Lagercrantz & Changeux, 2009). The final stages of the structural and functional sensorimotor development overlap in their timing, and their significant effects on the fetal behavioral outputs are clearly shown at late gestation. This may be related to their impact on the production of associative classes of neurons in deep proliferative zones (Dehay et al., 2015; Kalebic et al., 2019; Nowakowski et al., 2016; Popovitchenko & Rasin, 2017; Rakic, 1982).

Recent EEG findings sum up well the structural and functional aspects of early motivation clocks and the role of elicited and gradual developing inhibition along gestation. EEG delta bursts are apparent early in embryonic life before the cortical plate folds into six layers. While delta bursts are well accepted as driven by spontaneous activity, there are observations that indicate that these types of wave bursts are driven by signals from the sensory periphery, thus suggesting that the early random behaviors are generated by early formed organs such as the spinal cord until the timing of the early cortical activity. The late, sensation-oriented behavioral outputs, which were termed by others "endogenous stimulation," induce further development by the feedback of behaviors to the auditory and visual developing regions (Ackman & Crair, 2014; Huberman et al., 2008; Tritsch et al., 2007; Wang et al., 2015). Only later in gestation, the alpha wave, associated with awareness and awake-alert states, develops from the early delta waves (Goldstein Ferber et al., 2022).

It was found that movement-generating bursts in motor zones were followed by activity bursts in sensory zones (Inácio et al., 2016) and that this type of feedback affects the consolidation of thalamocortical spindle bursts and oscillations (Daw et al., 2007; Minlebaev et al., 2007, 2009, 2011). This consolidation phase was found to include the development of mounting inhibition by the onset of cortical inhibitory feedforward signaling (Hanganu-Opatz, 2010; Minlebaev et al., 2011).

Thus, early motivation clocks are based on mounting stop signals of the move-forward excitatory drive; they are shown in diversified inhibitory functions and evidenced in altering behavioral movements along gestation. These motivation clocks depend primarily on early transcriptional and molecular very early events, are shown mainly in peripheral signals generating stereotyped random movements during embryonic life, and are followed by the development of the sensorimotor regions and their feedback signaling to the thalamic projections. This is shown in sensation-oriented behaviors during late gestation. Contrary to the thought that the move-forward drive increases with gestation, this chapter shows that inhibitory factors are those which increase in healthy gestation, affording maturation of the early motivation clocks and their balance to the degree of fetal executed and repeated intentional behaviors.

References

Ackman, J. B., & Crair, M. C. (2014). Role of emergent neural activity in visual map development. *Current Opinion in Neurobiology, 24*, 166–175. https://doi.org/10.1016/j.conb.2013.11.011

Afif, A., Trouillas, J., & Mertens, P. (2015). Development of the sensorimotor cortex in the human fetus: A morphological description. *Surgical and Radiologic Anatomy, 37*(2), 153–160. https://doi.org/10.1007/s00276-014-1332-4

Arafa, K., & Emara, M. (2020). Insights about circadian clock and molecular pathogenesis in gliomas. *Frontiers in Oncology, 10*. https://doi.org/10.3389/fonc.2020.00199

Burdon, C., Mann, C., Cindrova-Davies, T., Ferguson-Smith, A. C., & Burton, G. J. (2007). Oxidative stress and the induction of cyclooxygenase enzymes and apoptosis in the murine placenta. *Placenta, 28*(7), 724–733. https://doi.org/10.1016/j.placenta.2006.12.001

Bystron, I., Blakemore, C., & Rakic, P. (2008). Development of the human cerebral cortex: Boulder committee revisited. *Nature Reviews Neuroscience, 9*(2), 110–122. https://doi.org/10.1038/nrn2252

Daw, M. I., Ashby, M. C., & Isaac, J. T. R. (2007). Coordinated developmental recruitment of latent fast spiking interneurons in layer IV barrel cortex. *Nature Neuroscience, 10*(4), 453–461. https://doi.org/10.1038/nn1866

Dehay, C., Kennedy, H., & Kosik, K. S. (2015). The outer subventricular zone and primate-specific cortical complexification. *Neuron, 85*(4), 683–694. https://doi.org/10.1016/j.neuron.2014.12.060

Donovan, T., Dunn, K., Penman, A., Young, R. J., & Reid, V. M. (2020). Fetal eye movements in response to a visual stimulus. *Brain and Behavior, 10*(8), e01676. https://doi.org/10.1002/brb3.1676

Einspieler, C., Prayer, D., & Marschik, P. B. (2021). Fetal movements: The origin of human behaviour. *Developmental Medicine & Child Neurology, 63*(10), 1142–1148. https://doi.org/10.1111/dmcn.14918

Fagard, J., Esseily, R., Jacquey, L., O'Regan, K., & Somogyi, E. (2018). Fetal origin of sensorimotor behavior. *Frontiers in Neurorobotics, 12*. https://doi.org/10.3389/fnbot.2018.00023

Ferber, S. G., Geva, R., & Weller, A. (2023). When the mind comes to live inside the body: The ontogeny of the perceptual control clock. *Current Neuropharmacology, 21*(1), 13–21. https://doi.org/10.2174/1570159X20666220411095508

Ferber, S. G., Weller, A., & Soreq, H. (2024). Boltzmann's theorem revisited: Inaccurate time-to-action clocks in affective disorders. *Current Neuropharmacology, 22*(11), 1762–1777. https://doi.org/10.2174/1570159X22666240315100326

Goldstein Ferber, S., Weller, A., Ben-Shachar, M., Klinger, G., & Geva, R. (2022). Development of the ontogenetic self-regulation clock. *International Journal of Molecular Sciences, 23*(2), Article 2. https://doi.org/10.3390/ijms23020993

Grunau, R. E. (2020). Personal perspectives: Infant pain—A multidisciplinary journey. *Paediatric and Neonatal Pain, 2*(2), 50–57. https://doi.org/10.1002/pne2.12017

Hanganu-Opatz, I. L. (2010). Between molecules and experience: Role of early patterns of coordinated activity for the development of cortical maps and sensory abilities. *Brain Research Reviews, 64*(1), 160–176. https://doi.org/10.1016/j.brainresrev.2010.03.005

Huberman, A. D., Feller, M. B., & Chapman, B. (2008). Mechanisms underlying development of visual maps and receptive fields. *Annual Review of Neuroscience, 31*, 479–509. https://doi.org/10.1146/annurev.neuro.31.060407.125533

Inácio, A. R., Nasretdinov, A., Lebedeva, J., & Khazipov, R. (2016). Sensory feedback synchronizes motor and sensory neuronal networks in the neonatal rat spinal cord. *Nature Communications, 7*(1), 13060. https://doi.org/10.1038/ncomms13060

Kalebic, N., Gilardi, C., Stepien, B., Wilsch-Bräuninger, M., Long, K. R., Namba, T., Florio, M., Langen, B., Lombardot, B., Shevchenko, A., Kilimann, M. W., Kawasaki, H., Wimberger, P., & Huttner, W. B. (2019). Neocortical expansion due to increased proliferation of basal progenitors is linked to changes in their morphology. *Cell Stem Cell, 24*(4), 535–550.e9. https://doi.org/10.1016/j.stem.2019.02.017

Komiya, Y., & Runnels, L. W. (2017). Chapter 28—Magnesium and embryonic development. In J. F. Collins (Ed.), *Molecular, genetic, and nutritional aspects of major and trace minerals* (pp. 343–351). Academic Press. https://doi.org/10.1016/B978-0-12-802168-2.00028-2

Kostović, I., Radoš, M., Kostović-Srzentić, M., & Krsnik, Ž. (2021). Fundamentals of the development of connectivity in the human fetal brain in late gestation: From 24 weeks gestational age to term. *Journal of Neuropathology & Experimental Neurology, 80*(5), 393–414. https://doi.org/10.1093/jnen/nlab024

Lagercrantz, H., & Changeux, J.-P. (2009). The emergence of human consciousness: From fetal to neonatal life. *Pediatric Research, 65*(3), 255–260. https://doi.org/10.1203/PDR.0b013e3181973b0d

McKinstry, R. C., Mathur, A., Miller, J. H., Ozcan, A., Snyder, A. Z., Schefft, G. L., Almli, C. R., Shiran, S. I., Conturo, T. E., & Neil, J. J. (2002). Radial organization of developing preterm human cerebral cortex revealed by non-invasive water diffusion anisotropy MRI. *Cerebral Cortex, 12*(12), 1237–1243. https://doi.org/10.1093/cercor/12.12.1237

Minlebaev, M., Ben-Ari, Y., & Khazipov, R. (2007). Network mechanisms of spindle-burst oscillations in the neonatal rat barrel cortex in vivo. *Journal of Neurophysiology, 97*(1), 692–700. https://doi.org/10.1152/jn.00759.2006

Minlebaev, M., Ben-Ari, Y., & Khazipov, R. (2009). NMDA receptors pattern early activity in the developing barrel cortex in vivo. *Cerebral Cortex, 19*(3), 688–696. https://doi.org/10.1093/cercor/bhn115

Minlebaev, M., Colonnese, M., Tsintsadze, T., Sirota, A., & Khazipov, R. (2011). Early gamma oscillations synchronize developing thalamus and cortex. *Science, 334*(6053), 226–229. https://doi.org/10.1126/science.1210574

Nowakowski, T. J., Pollen, A. A., Sandoval-Espinosa, C., & Kriegstein, A. R. (2016). Transformation of the radial glia scaffold demarcates two stages of human cerebral cortex development. *Neuron, 91*(6), 1219–1227. https://doi.org/10.1016/j.neuron.2016.09.005

Popovitchenko, T., & Rasin, M.-R. (2017). Transcriptional and post-transcriptional mechanisms of the development of neocortical lamination. *Frontiers in Neuroanatomy, 11*. https://doi.org/10.3389/fnana.2017.00102

Rakic, P. (1982). Early developmental events: Cell lineages, acquisition of neuronal positions, and areal and laminar development. *Neurosciences Research Program Bulletin, 20*(4), 439–451.

Rogers, C. D., Moody, S. A., & Casey, E. S. (2009). Neural induction and factors that stabilize a neural fate. *Birth Defects Research Part C: Embryo Today: Reviews, 87*(3), 249–262. https://doi.org/10.1002/bdrc.20157

Tritsch, N. X., Yi, E., Gale, J. E., Glowatzki, E., & Bergles, D. E. (2007). The origin of spontaneous activity in the developing auditory system. *Nature, 450*(7166), 50–55. https://doi.org/10.1038/nature06233

van der Veen, D. R., & Gerkema, M. P. (2017). Unmasking ultradian rhythms in gene expression. *The FASEB Journal, 31*(2), 743–750. https://doi.org/10.1096/fj.201600872R

Wang, H. C., Lin, C.-C., Chong, R., Zhang-Hooks, Y., Agarwal, A., Ellis-Davies, G., Rock, J., & Bergles, D. E. (2015). Spontaneous activity of Cochlear hair cells triggered by fluid secretion mechanism in adjacent support cells. *Cell, 163*(6), 1348–1359. https://doi.org/10.1016/j.cell.2015.10.070

Wasserman, T., & Wasserman, L. (2020). *Motivation, effort, and the neural network model*. Springer. https://doi.org/10.1007/978-3-030-58724-6

Wilkinson, M., Kane, T., Wang, R., & Takahashi, E. (2017). Migration pathways of thalamic neurons and development of Thalamocortical connections in humans revealed by diffusion MR Tractography. *Cerebral Cortex, 27*(12), 5683–5695. https://doi.org/10.1093/cercor/bhw339

Xu, G., Takahashi, E., Folkerth, R. D., Haynes, R. L., Volpe, J. J., Grant, P. E., & Kinney, H. C. (2014). Radial coherence of diffusion Tractography in the cerebral white matter of the human fetus: Neuroanatomic insights. *Cerebral Cortex, 24*(3), 579–592. https://doi.org/10.1093/cercor/bhs330

Chapter 9
The Clocks of Reward: Rhythms of Switches in Excitation Trigger Pleasure-Like Behaviors

Abstract Fetal early experiences of reward are reflected in clocks of performing a single pleasure-like incident, as well as in clocks of their reoccurrences in trajectories along gestation. Unlike previous descriptions of rewarding experiences, this chapter will describe the development of the sense of reward as corresponding to the rhythms of arousal buildup and return to excitation-inhibition balance. These clocks include known structures involved in the experience of reward, but they will be presented as dependent on the earlier molecular formations and the coordination of these formations between various levels. Those levels range from genetics, ion channels and cellular balance, neurotransmitter development, up to a complex timely intermixing quicker and slower rates of reward-processing regional growth. This chapter outlines the granular components for the differential time-sensitive rates of growth within the corticostriatal-pallidum-thalamic circuit, responsible for the origin and control of pleasure-like fetal behaviors. As such, this chapter outlines the role of GABA, glutamate, and acetylcholine and their switches on the excitatory-inhibitory axis through mid-gestation. Findings that at mid-gestation dopaminergic receptors in the immature prefrontal cortex are involved in the development of associative neurons, especially neurons involved in working memory, are presented. The structural advance of the regional network responsible for pleasure-like experiences and behaviors is presented here as based on early gestational functions of ionic regulation, especially chloride, sodium, and potassium, which in turn is governed by ultradian oscillations of clock genes for timely ion transport and cell osmotic regulation. This chapter also shows the assisting role of glial cells during early development as generated by ionic interplay, affecting timely development of choroid plexus macrophages. The time window of mid-gestation, associated with the borderline of viability and the onset of pleasure-like fetal behaviors, is suggested as the time of convergence between earlier molecular building blocks and structural-functional network establishment in the corticostriatal-pallidum-thalamic circuit. It is concluded that the clocks of reward reflect the time-sensitive process of ultradian on-off switches in excitation levels governed by clock genes, leading towards the onset of experience-based pruning routes, involving matched excitatory and inhibitory synapses to enable time-limited mounting of arousal and timely return to balance for any executed pleasure-like fetal behavior from mid-gestation onwards.

S. Goldstein Ferber, *Very Early Clocks*,
https://doi.org/10.1007/978-3-031-89603-3_9

The earliest sense of reward is expressed by the earliest developing sense, touch (Ferber et al., 2002), in the form of self-touching from mid-gestation onwards (Einspieler et al., 2021; Fagard et al., 2018). Early rewarding experience, especially when signaling involves sensorimotor and associative neurons as well, prepares the fetus for human communication and paves the way to shaping early learning capacities and executive functions. Yet important questions arise at this point. Is an early rewarding experience excitatory or inhibitory? Does it represent the return to balance after mounting excitatory inputs? And what characterizes an early rewarding experience? Is early rewarding experience characterized by patterns that rather represent the entire cycle of mounting arousal through the return to balance? Is this the developmental sequence shown in structural and functional molecular preparations during earlier stages of gestation, even embryonic?

The human striatum, the main brain region known to be involved in rewarding experiences, is composed of caudate, putamen, and the ventral stratum nuclei. The nucleus accumbens is located in ventral striatum. The striatum has been reported to emerge as early as 15 weeks of gestation. Migration of cells from the ganglionic eminence to one of the striatum's growing entities, the putamen, occurs until 17 weeks of gestation. Formation of neurons occurs at this timing, and by 25–27 weeks of gestation, the globus pallidus is divided to two parts, and the striatum's structural forms are then similar to the adult configuration (Hu et al., 2012). Therefore, it is not surprising that after mid-gestation pleasure-like repeated self-touching has been found by 3D ultrasound recordings (Donovan et al., 2020; Einspieler et al., 2021). These pleasure-like behaviors occur concomitantly with the maturation of sleep states, sucking behaviors and the coupling of heart rate, sleep states, and motor movements in fetal development (Ferber et al., 2023a). However, in this chapter, the clocks of reward are presented as produced by complex molecular, regional, and inter-regional development, beyond the maturation of the striatum.

Unlike previous descriptions of rewarding experiences, this chapter will describe the development of the sense of reward as corresponding to the rhythms of arousal buildup and return to excitation-inhibition balance, developing from the dominance of excitatory factors in the embryo, up to the emergence of excitation-inhibition balance at the junction of mid-gestation, and further developing beyond, towards the timing of expected term birth. This chapter will argue that the infrastructure for the healthy progress of the corticostriatal-pallidum-thalamic circuit genesis is based on complex synaptogenesis along this circuit, while genetic control fuels this process. Clock genes will be presented as controlling the rhythms of changes in ion interplay within the ion channels at different stages of gestation. Dopamine, acetylcholine, and glutamate neurotransmitters along with their receptors' availability will be presented as the end point of the molecular basis producing pleasure-like behaviors. The clocks of reward will be described as developmental trajectories and as underlying a single behavioral pleasure-like incident as well. The developmental trajectories will be presented as reoccurrences of a single pleasure-like behavior.

As early embryonic life is highly excitatory, these achievements of pleasure-like behaviors are apparent not earlier than the borderline of viability, when painful experiences are detectable too (Grunau, 2020). At this timing, the synchronization between the early developing sympathetic system and the later developing parasympathetic system, both originating from the brainstem, has been reported (Goldstein Ferber et al., 2022). This synchronization serves as another marker of the repeated mounting arousal towards the return to balance, coordinating fetal structural, functional, and behavioral developmental status at a given incident. This type of incidents is in turn observable and increases its frequency as gestation progresses (Einspieler et al., 2021; Goldstein Ferber et al., 2022).

How does the early rewarding experience start, and what are the neural and biochemical seeds of its entire cycle, ending in return to balance, when a behavioral action is fully executed? In the human embryo, the striatum, originating from the telencephalon, is divided into its compartments that are relatively close in anatomical architecture but functionally distinct: the dorsal and the ventral striatal compartments. The mature striatum structure is composed of massive innervation of excitatory glutamatergic afferents from the cortex and thalamus, and its developmental trajectories during gestation involve the establishment of time-dependent receptor availability and accurate release of dopamine, acetylcholine, and glutamate (Herlenius & Lagercrantz, 2004). The striatum is very active in the immature brain as a highly metabolic functioning region, and it possesses a high concentration of excitatory receptors, thus being vulnerable to injuries and deviations from the time-preprogrammed developmental trajectories aimed at gradually generating excitation-inhibition balance conditions (Ten Donkelaar et al., 2023; Tian et al., 2022).

Furthermore, in the embryo, early developing neural epithelial cells produce other cells in the ventricular region, which in turn send projections to the subventricular region of the ganglionic eminence, thus shaping the radial glia, the source of the spiny projection neurons (SPNs) cells, which are GABAergic in the striatum, and differentiate into two distinct subpopulations of progenitors during striatal neurogenesis, apical and then basal.

A recent study (Kelly et al., 2018; Lebouc et al., 2020) identified two stages: first, during embryonic early development, restricted to production of apical cells, which in turn give rise later to the development of striosomal SPNs, and, second, a phase at mid-gestation, a longer phase following intermediate differentiation of the basal cells generating almost all the matrix of the SPNs and the onset of striatal genesis with some overlaps in time between the two phases. Another impressive study (Tinterri et al., 2018) found that the migration profile of these cells and their early specification occurs early in the embryonic striatum. One example of early arousal inhibitors is the early B-cell factor 1 (Ebf1) which appears to inactivate this process early in embryogenesis, leading to its complexity by producing altered dSPN/iSPN (direct and indirect) intermixing. This shows that the cell speciation is dependent on timing and on interacting activation and inactivation of molecular factors, corresponding to early experiences of excitation-inhibition balance (Tian et al., 2022). The inactivating factors form the early incidents of the return to balance from high excitatory conditions and serve as the seeds of the entire single incident of the rewarding experience cycle, which emerges as such only after mid-gestation but develops from this early basis.

To explain the neural basis for the early rewarding experience such as a behavioral incident resulting from mounting striatal dopaminergic excitation followed by the developing cortical inhibitory signals toward achieving excitation-inhibition balance, it is noted that this is a typical early developmental pattern of the corticostriatal-pallidum-thalamic circuit. This circuit is a fundamental neuronal circuit for cortical regulation of complex sensory, motor, limbic, and associative functions modulated by the amygdala, subthalamus, and mesencephalic tegmental nuclei inputs (Goldman-Rakic, 1981; Graybiel & Ragsdale, 1980; Schmahmann & Pandya, 2009), from mid-gestation onwards. This is shown at the time window in fetal progressing control of behavioral movements, along a cycle of mounting excitation up to the sinusoidal pattern of return to balance. This is considered here as the behavioral presentation of the earliest sense of reward. By 24–27 weeks of gestation, which is also the timing of complex fetal behavioral onset, the corticostriatal pathways are vulnerable to injuries more than earlier and later, and this window of vulnerability occurs when the corticostriatal terminal fibers embrace the striatal cell islands. At 28–38 weeks of gestation, the fibers in this circuit reach their maximal growth while showing a transient form which becomes more evenly distributed later in gestation when the putamen and the caudate of the circuit mosaic are fully innervated (Kostović et al., 2021).

Corticostriatal axonal connectivity is established after mid-gestation and therefore corticostriatal innervation during embryonic stages of development, before synaptogenesis, occurs without cortical modulation of input and outputs. The striatum develops rapidly during the second trimester of gestation with regional complex time-dependent trajectories. The inputs to the corticostriatal circuit are vastly heterogeneous, and this nature makes their developing synapses vulnerable during the time window of mid-gestation with implications for neurological diseases later in life (Kuo & Liu, 2019). Neuron migration in the human striatum continues at least from 21 to 34 weeks of gestation with the total number of neurons increasing. After 34 weeks of gestation, the size of the ganglionic eminence decreases indicating a decrease in proliferation (Kostović et al., 2021).

The gross volume development of the striatum occurs concomitantly with other neurogenetic events: the cells from the ganglionic eminence run from the ventricular surface towards the caudate and the putamen increasing the total number of the striatal active neurons. This period of development is characterized by the massive invasion of projection fibers into the corticostriatal junction, and those long axonal pathways increase in number and size, thus altering and maturing corticostriatal functions including the number of active synapses and the gross size of the striatum early in the third semester of gestation (Tian et al., 2022). The rate of striatal growth is fast during 14–22 weeks of gestation, determining cell proliferation and migration, while the development of the cerebral cortex neurons is generated in the ventricular and subventricular zones, migrating to the pial surface (Kostović et al., 2019).

Research showed that projections from the cortex were organized according to medial-lateral and rostral-caudal gradients along the striatal nuclei (Jarbo & Verstynen, 2015). In another study, it was found that the medial and the lateral subdivisions of the caudate nucleus head and body grew faster than the other regions, which were interconnected with the medial prefrontal cortex during the later stages of gestation (Graff-Radford et al., 2017). The dorsal and the ventral regions of the putamen grew faster than the other regions receiving projections from premotor cortex, motor cortex, and the dorsolateral prefrontal cortex involved in motor actions (Jarbo & Verstynen, 2015). Thus, the regions of the caudate, involved in associative and cognitive developing functions, grew faster than those involved in motor functions. The opposite has been recognized in the putamen, that is, regions involved in motor functions grew faster than others (Tian et al., 2022).

The complex of rapid vs. slower growth of striatal structures in diverse curves over time, is evident in findings that show a junction of changes around day 20 of gestation, in both linear and quadratic patterns. This structural complex reflects yet a functional facet of the synaptic infrastructure developing balance between speeding and decelerating striatal developmental rhythms. These rhythms develop towards trajectories which produce repetitions of single early rewarding incidents by repeated returns to balance and elimination of excitotoxicity from mid-gestation onwards during healthy gestation. Notably, differences in excitation and inhibition, that are shown in the postsynaptic areas, become more stable and accurate in functionality from mid-gestation onwards, while the specification of the involved cells is acquired earlier during development. Excitatory contacts maintain asymmetry with the electron-dense postsynaptic density, opposing presynaptic active spaces. In contrast, inhibitory signaling appears in a symmetric manner with postsynaptic spaces clustered on the synaptic cleft without significant postsynaptic density (Boeckers, 2006; Okabe, 2007; Südhof, 2021; Tyagarajan & Fritschy, 2010).

Beyond genetic control of the excitatory-inhibitory ratio, compensatory mechanisms are apparent in the attempt of the organism to reach a set point for the complex feedback mechanisms of synaptogenesis to the genetic level, a type of feedback which is time-dependent, may reach different ratios at different times of gestation, and is suggested to underly a complex, nonlinear, slope of increase in behavioral pleasure-like incident repetitions over the time of gestation (Gatto & Broadie, 2010). Intrinsic excitability of SPNs progressively decreases as gestation progresses, affected by hyperpolarization of their resting membrane potential and prolonged spike latency to firing. This implies elevated action potential thresholds as gestation progresses closer to birth.

Others related this to reorganization to the more accurate function of extracellular potassium ion channels (Krajeski et al., 2019), as the crucial regulator of striatal dopamine release. This suggests growing excitation-inhibition balance along the progress of gestation and that at least to the extent of the N^+K^+ pump function, the increase in resting potential of the cell serves as a developing basis for accomplishing the return to balance when a single behavioral action is fully executed. By feedback mechanisms, this contributes to the establishment of excitation-inhibition balance over the time of gestation as a necessary component for the development of reward-related structures, beyond its general importance for healthy maturation of the organism (Dutta et al., 2023; Lebouc et al., 2020).

Thus, postsynaptic conditions underlying a single early rewarding experience predict its repetitions and the increase in frequency of this type of incidents. As this type of incidents is typically created by timely excitation and then return to balance, the question that arises here is: what are the rhythms that this structural-functional advance is responsive to? How does the fetal organism know not just how but also when to excite and when to suppress and inhibit?

Clock genes are designed for on-off switches of molecular events through their heterodimerizing patterns, their modulation of Ca^{2+} entries to the cell, and their capacity to convert unstable molecules to stable ones (Hansen et al., 2020; Salihu et al., 2021; van der Veen & Gerkema, 2017). The on-off impact of clock genes in early gestation has been found to span the transcription of the whole genome (Cox & Takahashi, 2019). The structural advance of corticostriatal-pallidum-thalamic circuit described here is based on early gestational functions of ionic regulation, especially chloride, sodium, and potassium, which in turn is governed by ultradian oscillations of clock genes for timely ion transport and cell osmotic regulation. These molecular events trigger organismal on-off phases of excitation, thus dictating the rate of structural volume growth. The rhythms governed by clock genes have implications for neurodevelopmental risks. It has been recently suggested that genetic control of synaptogenesis may shift from normal rhythms of development, thus affecting structural configuration. Factors causing such deviations on the vector of time include the selective loss of either excitatory or inhibitory synapses without compensatory changes. Such genetic conditions contribute to the formation and maintenance of one class of synapse without being relative to the other, thus creating an excitation-inhibition ratio which is not time appropriate for gestational age (Boeckers, 2006; Okabe, 2007; Tyagarajan & Fritschy, 2010).

Thus, the level and accuracy of the function of the given structure, matching remaining excitatory and inhibitory synapses on a time-dependent manner during synaptogenesis, at least with regard to the progress from hyperexcitability in the embryo toward excitation-inhibition balance from mid-gestation onward, is crucial for governing the development of neural circuits underlying the development of rewarding experiences. The infrastructure of structural-functional reciprocal effects, under genetic control for the development of synaptogenesis, is mainly the development of neurotransmitters and the availability of their related receptors.

In particular, dopamine acetylcholine and glutamate predominantly govern the synaptogenesis and massive innervation in the corticostriatal-pallidum-thalamic circuit, thus triggering behavioral progress in fetal repetitions of early rewarding experiences such as self-touching. This echoes in other developing regional circuitries too, as observed, for example, in the growing complexity of motor movements and their coupling with heart rate and discriminated sleep states from mid-gestation onwards (Ferber et al., 2023a). Dopaminergic neurons are apparent in the human embryo as early as 6–8 weeks of gestation with a high turnover, reflecting high excitability in early gestation, and the potential for early experience of reward derived from arousal which is not yet coupled with inhibitory factors. This occurs while the organization of behavioral output is not yet apparent, and the sense of early reward may also be abbreviated and incidental. Upon the resolution of the cortical subplate and the achievement of a six-layered immature cortex, dopaminergic receptors were found in extremely high levels in the pallidum (Salihu et al., 2021), a location related to reward processing. At this stage of development, from mid-gestation onward, these dopaminergic receptors, located in the immature prefrontal cortex, were found to be involved in the development of associative neurons and especially in neurons involved in working memory. At this period of gestation, a sense of early rewarding experience is likely processed by an established network on the corticostriatal-pallidus-thalamic circuit which may involve the entire experience of building up arousal and then return of excitation inhibition balance. Interferences with prefrontal dopamine through feedback to the genetic level at different early and later stages of gestation were found to be associated with deviations from healthy development, including gaps in the signaling between growing associative neurons and the sensorimotor projections at this timing (Diamond et al., 2004; Ferber et al., 2023a).

How does the dominance of excitation work in the embryo, and what is its contribution to the later more consolidated sense of rewarding experience? Adrenaline in the embryo has an inhibitory function on respiratory regulation by the brainstem, and GABA has an excitatory signaling up to mid-gestation. During early development, the Cl^- concentration is high in nerve cells. When GABA, as an excitatory agent in early embryonic phases, opens the Cl^- channels, a depolarization and excitatory conditions occur. During later stages of gestation and further maturation, Cl^- concentration decreases, which turns GABA's role to inhibitory at mid-gestation, in which Cl^- moves out of the cell while the cell becomes hyperpolarized, representing the seeds of excitation-inhibition balance. At early stages of gestation, AMPA receptors and the NMDA channels are blocked by Mg^+ at resting membrane conditions, but this blockade is moved by GABA signaling for the early contribution of promoting synaptic plasticity and the establishment of neural networks. GABA-stimulated upregulation of the potassium-chloride cotransporter KCC2 was suggested as the basis for synapse switch from excitatory to inhibitory (Kriegstein & Owens, 2001) later in gestation.

This complex reflects a whole brain underlying mechanism of the early experience of reward by accomplishing the entire process from the arousal buildup and avoidance from flaccid conditions up to the return to balance. From the same view, this complex of the rewarding experience, with a balanced basis at later stages of gestation, is shown in acetylcholine's role, as it is heavily involved in cortical activation, including reward and pain, counterbalancing the excitatory dopaminergic effects (Cooper et al., 2003). Acetylcholine is also accepted as interacting with myelinating glia and responsible for electrogenesis and firing rate of neurons with developmental progress of switches from excessive bursts in the embryo to increased propagation delays and membrane resting potentials from mid-gestation onwards. Nicotinic acetylcholine receptors activate a large extent of silent synapses producing a calcium-induced release with high excitatory reaction and behavioral arousal, which may be time-appropriate in the embryo but could run up to excitotoxicity and the risk of apoptosis from mid-gestation onwards. At optimal levels, acetylcholine regulates dopamine for major development adjustments in early cortical organization. The glutamatergic synapses form exclusively dendritic spines after stabilization of input (Lohmann & Bonhoeffer, 2008).

In contrast, GABAergic synapses do not form on spines but rather develop on dendritic shafts, nerve cell in somatic organs, and axon initial segments. These inhibitory synapses arise from predecessor axonal-dendrite connectivity which are established in a complex manner early in gestation but without transferring activity from the axon or dendrite at the earliest stages of gestation. Glutamate is the major excitatory agent of development from mid-gestation onwards, excessive glutamate release leads to C^{2+} efflux, and may include the risk of apoptosis. Deviations from optimal ranges of NMDA receptor activity, triggered by glutamate, can be life-threatening (Lohmann & Bonhoeffer, 2008).

That is, the growing signaling towards the buildup of rewarding arousal earlier than the midgestational junction may be risky to the entire organism beyond any genetic control. This requires growing resting membrane potentials, which appear later in gestation, for the maturation of the clocks of reward and healthy continuation of gestation, especially given the excitatory signaling that gestation reaches during the advanced phases closer to term birth. The neurotransmitters and the availability of receptors are the last point of control before a pleasure-like incident of behavioral output is executed as well as the point of feedback to genetic, metabolic, cellular, network, and regional levels (Ferber et al., 2023b, 2024). These are the organismal conditions affording the progress toward relative maturation of pleasure-like behaviors such as repeated self-touching.

In sum, pleasure-like behaviors and the sense of rewarding experience leading to repetition of such behaviors are dependent on the timing along gestation: early in gestation rewarding experiences may be related to overexcitation eliciting stereotyped random embryonic movements, and around mid-gestation they may be related to the emerging excitation-inhibition balance, while close to term they may be again based on excitation buildup, shown in more frequent repetitions of complex behaviors. Thus, the implied early experiences of reward are reflected in clocks of a single incident as well as in clocks of developmental trajectories along gestation. Indeed, these clocks include known structures involved in the experience of reward, but they are dependent on the cooperating function of the entire immature brain levels, from ion channels and cellular formation up to complex timely intermixing quicker and slower rates of reward-processing regional growth. That is, the clocks of repeated pleasure-like behavior generation represent a composition of the genetic-level crosstalk with the time-appropriate developing infrastructure of synaptogenesis and neurotransmitters. As such, these clocks reflect the time-sensitive process of ultradian on-off switches in excitation levels governed by clock genes, leading towards the onset of experience-based pruning routes, involving matched excitatory and inhibitory synapses to enable time-limited mounting of arousal and timely return to balance for any executed pleasure-like fetal behavior from mid-gestation onwards.

References

Boeckers, T. M. (2006). The postsynaptic density. *Cell and Tissue Research, 326*(2), 409–422. https://doi.org/10.1007/s00441-006-0274-5

Cooper, J. R., Bloom, F. E., & Roth, R. H. (2003). *The Biochemical Basis of Neuropharmacology*. Oxford University Press.

Cox, K. H., & Takahashi, J. S. (2019). *Circadian clock genes and the transcriptional architecture of the clock mechanism*. https://doi.org/10.1530/JME-19-0153

Diamond, A., Briand, L., Fossella, J., & Gehlbach, L. (2004). Genetic and Neurochemical Modulation of Prefrontal Cognitive Functions in Children. *American Journal of Psychiatry, 161*(1), 125–132. https://doi.org/10.1176/appi.ajp.161.1.125

Donovan, T., Dunn, K., Penman, A., Young, R. J., & Reid, V. M. (2020). Fetal eye movements in response to a visual stimulus. *Brain and Behavior, 10*(8), e01676. https://doi.org/10.1002/brb3.1676

Dutta, S., Iyer, K. K., Vanhatalo, S., Breakspear, M., & Roberts, J. A. (2023). Mechanisms underlying pathological cortical bursts during metabolic depletion. *Nature Communications, 14*(1), 4792. https://doi.org/10.1038/s41467-023-40437-0

Einspieler, C., Prayer, D., & Marschik, P. B. (2021). Fetal movements: The origin of human behaviour. *Developmental Medicine & Child Neurology, 63*(10), 1142–1148. https://doi.org/10.1111/dmcn.14918

Fagard, J., Esseily, R., Jacquey, L., O'Regan, K., & Somogyi, E. (2018). Fetal Origin of Sensorimotor Behavior. *Frontiers in Neurorobotics, 12*. https://doi.org/10.3389/fnbot.2018.00023

Ferber, S. G., Geva, R., & Weller, A. (2023a). When the Mind Comes to Live Inside the Body: The Ontogeny of the Perceptual Control Clock. *Current Neuropharmacology, 21*(1), 13–21. https://doi.org/10.2174/1570159X20666220411095508

Ferber, S. G., Laudon, M., Kuint, J., Weller, A., & Zisapel, N. (2002). Massage therapy by mothers enhances the adjustment of circadian rhythms to the nocturnal period in full-term infants. *Journal of Developmental & Behavioral Pediatrics, 23*(6), 410.

Ferber, S. G., Weller, A., & Soreq, H. (2023b). Control systems theory revisited: New insights on the brain clocks of time-to-action. *Frontiers in Neuroscience, 17*. https://doi.org/10.3389/fnins.2023.1171765

Ferber, S. G., Weller, A., & Soreq, H. (2024). Boltzmann's theorem revisited: inaccurate time-to-action clocks in affective disorders. *Current Neuropharmacology, 22*(11), 1762–1777. https://doi.org/10.2174/1570159X22666240315100326

Gatto, C. L., & Broadie, K. (2010). Genetic controls balancing excitatory and inhibitory synaptogenesis in neurodevelopmental disorder models. *Frontiers in Synaptic Neuroscience, 2*. https://doi.org/10.3389/fnsyn.2010.00004

Goldman-Rakic, P. S. (1981). Prenatal formation of cortical input and development of cytoarchitectonic compartments in the neostriatum of the rhesus monkey. *Journal of Neuroscience, 1*(7), 721–735. https://doi.org/10.1523/JNEUROSCI.01-07-00721.1981

Goldstein Ferber, S., Weller, A., Ben-Shachar, M., Klinger, G., & Geva, R. (2022). Development of the ontogenetic self-regulation clock. *International Journal of Molecular Sciences, 23*(2), Article 2. https://doi.org/10.3390/ijms23020993

Graff-Radford, J., Williams, L., Jones, D. T., & Benarroch, E. E. (2017). Caudate nucleus as a component of networks controlling behavior. *Neurology, 89*(21), 2192–2197. https://doi.org/10.1212/WNL.0000000000004680

Graybiel, A. M., & Ragsdale, C. W. (1980). Clumping of acetylcholinesterase activity in the developing striatum of the human fetus and young infant. *Proceedings of the National Academy of Sciences, 77*(2), 1214–1218. https://doi.org/10.1073/pnas.77.2.1214

Grunau, R. E. (2020). Personal perspectives: Infant pain—A multidisciplinary journey. *Paediatric and Neonatal Pain, 2*(2), 50–57. https://doi.org/10.1002/pne2.12017

Hansen, J. M., Jones, D. P., & Harris, C. (2020). The redox theory of development. *Antioxidants & Redox Signaling, 32*(10), 715–740. https://doi.org/10.1089/ars.2019.7976

Herlenius, E., & Lagercrantz, H. (2004). Development of neurotransmitter systems during critical periods. *Experimental Neurology, 190*, 8–21. https://doi.org/10.1016/j.expneurol.2004.03.027

Hu, Z., Rudd, J. A., & Fang, M. (2012). Development of the human corpus striatum and the presence of nNOS and 5-HT2A receptors. *The Anatomical Record, 295*(1), 127–131. https://doi.org/10.1002/ar.21497

Jarbo, K., & Verstynen, T. D. (2015). Converging structural and functional connectivity of orbitofrontal, dorsolateral prefrontal, and posterior parietal cortex in the human striatum. *Journal of Neuroscience, 35*(9), 3865–3878. https://doi.org/10.1523/JNEUROSCI.2636-14.2015

Kelly, S. M., Raudales, R., He, M., Lee, J. H., Kim, Y., Gibb, L. G., Wu, P., Matho, K., Osten, P., Graybiel, A. M., & Huang, Z. J. (2018). Radial glial lineage progression and differential intermediate progenitor amplification underlie striatal compartments and circuit organization. *Neuron, 99*(2), 345–361.e4. https://doi.org/10.1016/j.neuron.2018.06.021

Kostović, I., Radoš, M., Kostović-Srzentić, M., & Krsnik, Ž. (2021). Fundamentals of the development of connectivity in the human fetal brain in late gestation: From 24 weeks gestational age to term. *Journal of Neuropathology & Experimental Neurology, 80*(5), 393–414. https://doi.org/10.1093/jnen/nlab024

Kostović, I., Sedmak, G., & Judaš, M. (2019). Neural histology and neurogenesis of the human fetal and infant brain. *NeuroImage, 188*, 743–773. https://doi.org/10.1016/j.neuroimage.2018.12.043

Krajeski, R. N., Macey-Dare, A., van Heusden, F., Ebrahimjee, F., & Ellender, T. J. (2019). Dynamic postnatal development of the cellular and circuit properties of striatal D1 and D2 spiny projection neurons. *The Journal of Physiology, 597*(21), 5265–5293. https://doi.org/10.1113/JP278416

Kriegstein, A. R., & Owens, D. F. (2001). GABA may act as a self-limiting trophic factor at developing synapses. *Science's STKE, 2001*(95), pe1–pe1. https://doi.org/10.1126/stke.2001.95.pe1

Kuo, H.-Y., & Liu, F.-C. (2019). Synaptic wiring of corticostriatal circuits in basal ganglia: Insights into the pathogenesis of neuropsychiatric disorders. *eNeuro, 6*(3). https://doi.org/10.1523/ENEURO.0076-19.2019

Lebouc, M., Richard, Q., Garret, M., & Baufreton, J. (2020). Striatal circuit development and its alterations in Huntington's disease. *Neurobiology of Disease, 145*, 105076. https://doi.org/10.1016/j.nbd.2020.105076

Lohmann, C., & Bonhoeffer, T. (2008). A role for local calcium signaling in rapid synaptic partner selection by Dendritic Filopodia. *Neuron, 59*(2), 253–260. https://doi.org/10.1016/j.neuron.2008.05.025

Okabe, S. (2007). Molecular anatomy of the postsynaptic density. *Molecular and Cellular Neuroscience, 34*(4), 503–518. https://doi.org/10.1016/j.mcn.2007.01.006

Salihu, S., Meor Azlan, N. F., Josiah, S. S., Wu, Z., Wang, Y., & Zhang, J. (2021). Role of the cation-chloride-cotransporters in the circadian system. *Asian Journal of Pharmaceutical Sciences, 16*(5), 589–597. https://doi.org/10.1016/j.ajps.2020.10.003

Schmahmann, J. D., & Pandya, D. (2009). *Fiber Pathways of the Brain*. Oxford University Press.

Südhof, T. C. (2021). The cell biology of synapse formation. *Journal of Cell Biology, 220*(7), e202103052. https://doi.org/10.1083/jcb.202103052

Ten Donkelaar, H. J., Lammens, M., & Hori, A. (2023). *Clinical neuroembryology: Development and developmental disorders of the human central nervous system*. Springer. https://doi.org/10.1007/978-3-031-26098-8

Tian, M., Xu, F., Xia, Q., Tang, Y., Zhang, Z., Lin, X., Meng, H., Feng, L., & Liu, S. (2022). Morphological development of the human fetal striatum during the second trimester. *Cerebral Cortex, 32*(22), 5072–5082. https://doi.org/10.1093/cercor/bhab532

Tinterri, A., Menardy, F., Diana, M. A., Lokmane, L., Keita, M., Coulpier, F., Lemoine, S., Mailhes, C., Mathieu, B., Merchan-Sala, P., Campbell, K., Gyory, I., Grosschedl, R., Popa, D., & Garel, S. (2018). Active intermixing of indirect and direct neurons builds the striatal mosaic. *Nature Communications, 9*(1), 4725. https://doi.org/10.1038/s41467-018-07171-4

Tyagarajan, S. K., & Fritschy, J.-M. (2010). GABAA receptors, gephyrin and homeostatic synaptic plasticity. *The Journal of Physiology, 588*(1), 101–106. https://doi.org/10.1113/jphysiol.2009.178517

van der Veen, D. R., & Gerkema, M. P. (2017). Unmasking ultradian rhythms in gene expression. *The FASEB Journal, 31*(2), 743–750. https://doi.org/10.1096/fj.201600872R

Chapter 10
The Clocks of Executive Functions: Rhythms of Area- and Time-Specific Progenitors Induce Stimulus-Related Behaviors

Abstract When is the first executive function identified? What are the early molecular building blocks of this first occurrence? What are the rhythms of their development on the vector of time, and how do they relate to overt early behavior? How do the molecular rhythms dictate morphological development up to elicitation of the first executive function? This chapter will address these questions.

Recent views based on groundbreaking 3D, 4D ultrasonic recordings suggest that early executive functions are shown in human fetal behavior and that this rich repertoire does not occur abruptly when the newborn meets the out-uterine environment as previously thought. It is shown in this chapter that the genetic clocks of expression and suppression within the prefrontal-thalamic-hypothalamic-hippocampal-corticospinal tract network develop in a combinatorial time-sensitive manner into cortical ability to reach short states of awakening. Once these brief states have occurred, the alert state span may widen in later repetitions of this condition often described as a paradoxical alpha wave with the return of the alpha wave to the EEG pattern. Differential rates of growth in the different prefrontal growing systems occur in parallel with growth in lower regions coordinating prefrontal-thalamic-hypothalamic-hippocampal-corticospinal tract network. This chapter argues that cortical and subcortical regions develop as early structural volumes towards each other and their mutual embrace elicits the components of the alert state and more complex behavior phenomena later in gestation. These processes develop while residual physiological and cortical excitability, which follows the behavioral execution of a function, decreases from mid-gestation onward, showing more coordination of network wiring and feedback loops from the behavioral to the molecular level. While the molecular basis affording the occurrence of short alert states and aligned early fetal executive functions encompasses a time-sensitive mode of expression, mid-gestation is the time when genetic and cellular clocks are beginning to be shaped by experience. This suggests that at this time window any remaining differentiation starts to overlap with pruning processes. It is hypothesized that this transition around mid-gestation is afforded by changes in the earlier combinatorial balance between molecules which promote vs. those which restrict expression. Furthermore, this balance gradually changes on the genetic-driven level to respond to environmental inputs too, resembling the earlier shown capacity of

S. Goldstein Ferber, *Very Early Clocks*,
https://doi.org/10.1007/978-3-031-89603-3_10

clock genes. It is concluded that although the clocks of executive functions are based on early molecular building blocks, from mid-gestation onwards, this molecular basis interacts with the environment too, in a manner that any fetal acquisition of an executive function cannot be predicted by either the clocks or the environmental stimuli only.

Recently, the answer to the question whether the newborn acquires executive functions only after birth and not before has been unpacked, suggesting that there is a continuation of the executive functionality from mid-gestation to postnatal capacities. Recent recordings of fetal behaviors show that behaviors such as habituation and the orienting response, related previously to the postnatal times only, do not appear abruptly with the birth experience or with the exposure to the out-uterine environment. These behaviors are detectable from mid-gestation onward. Without the introductory phase during the fetal period, it is hard to understand the newborn's relatively wide range of executive repertoire.

Recent reports mention earlier suggestions that early fetal visuomotor development is predominantly controlled by gestational clocks and not by out-uterine experience (Donovan et al., 2020). In this chapter, the continuation from early embryonic genetic and molecular building blocks and their ongoing function along gestation are presented. These clockwise trajectories are presented with their time-sensitive aligned behavioral changes through the onset of early executive functions at mid-gestation, developing towards late gestation and birth. It is shown in this chapter that residual physiological and cortical excitation that follows the more organized behaviors at mid-gestation decreases along late gestation and towards birth. The perspective in this chapter is described by showing data on behaviors during gestation at different stages and timings in association with genetic electroencephalogram, MRI, and fMRI data collected at the same gestational timings, suggesting that this association reflects the very early clocks of executive functions.

This chapter will also show how the genetic clocks are part of healthy gestation and that they develop in a combinatorial timewise manner into cortical ability to reach short states of awakening, combined with stimulus-oriented fetal behavioral reactions detectable by EEG wave changes and sonographic 3D, 4D recordings. This chapter will show that very early executive functions originate from internal clock-driven molecular and cellular events during the embryonic phase and that this clock is designed to develop into an experience-shaped pattern too, not excluding the clock itself. This occurs from mid-gestation onwards, when the connectivity of the prefrontal-thalamic-hypothalamic-hippocampal-corticospinal tract network turns to coordinated polarization-depolarization rhythms and executive functions become part of detectable fetal behavior.

In general, although molecular components were traditionally related to later developed seizures and cortical pathologies (Spoto et al., 2021), in this chapter, these components are presented as working continuously along healthy gestation with complex diverse and different impacts at different timings of gestation for later production of more organized stimulus-related behaviors. These molecular components are further described here as contributing to regional volume development and inter-regional connectivity within the prefrontal-thalamic-hypothalamic-hippocampal-corticospinal tract network.

Until the 1980s, the fetus had been considered as undergoing a "developmental limbo" (Hopkins & Johnson, 2005). Although fetal attention towards a stimulus can only be intermittently detected, the phenomenon of transitory awake/awareness of the fetus following discrimination between active and quiet sleep at mid-gestation (Ferber et al., 2023) cannot be ignored. Recent findings showed that levels of anticipatory awareness are present in the fetus, developing from mid-gestation to a clear orienting response at 34 weeks of gestation (Myowa-Yamakoshi & Takeshita, 2006; Reid et al., 2017; Reissland et al., 2014). Groundbreaking experiments during the last decade showed that the fetus is an active participant in intra-uterine environment-driven interactions observed by 3D, 4D ultrasonic recordings, including perception of – and response to outside information. This is reflected in anticipatory self-touching behaviors (mouth opening occurs before the hand-to-mouth movement) [Reissland et al., 2014]) and eye movements (following an intrauterine light stimulus [Donovan et al., 2020]). These behaviors were originally included in the early orienting response, related in the past to the postnatal period and not earlier than that.

Part of the old view on the postnatal onset timing of the orienting response was based on difficulties in sleep state classifications and especially the difference between awake and non-awake states. These difficulties resulted also from the lack of data on the complex association between physiological arousal (e.g., heart rate), cortical arousal, and clear executive functions. In fetuses aged 240 days post-conception who reacted with a short transitory awake state to an intra-uterus light stimulus, this complex has been shown by residual cortical arousal in the form of continued eye movements, even though head turns were eliminated after the intra-uterine light stimulus stopped. Other findings showed that during the recording of the alpha wave, the best marker of visual information processing in awake states, heartbeat per minute extended above mean values, and this was associated with uterine contractions more than with the changes in spectral power. During the shifts from alpha to theta band, spectral edge frequency declined (Thaler et al., 2000). These data show that general physiological arousal includes residual excitatory components and that it is not reflecting identical clear alert states on the vector of time. It also shows that the residual components of arousal decrease with age (Jabbari, 2021) along with accompanying more organized behaviors during phases of alpha wave recordings (Einspieler et al., 2021; Fagard et al., 2018).

Cortical feedforward loops to subcortical regions progressing to the production of cortical prevailing commands over subcortical regions can be detected in changes in EEG waves from mid-gestation to early childhood. These include short awake states which follow a developmental trajectory of the alpha wave origination from the tracé-alternant discontinuous pattern of high-amplitude delta-theta activity bursts (Mizrahi & Hrachovy, 2015; Myers et al., 1997). True fetal executive functions, although they may be short, can be detected in awake states with wide eye opening and aligned complex behaviors such as eye movements with head turning or hand to mouth, showing brief alpha waves. Once these brief states have occurred, the alert state span may widen in later repetitions of this condition often described as a paradoxical alpha wave with the return the alpha wave to the EEG pattern (Britton et al., 2016; Einspieler et al., 2021; Fagard et al., 2018).

Alpha waves have been widely accepted as representing visual processing of outside information and awake/alert states, be it a significant representation or a fragmentary one. From mid-gestation onwards, alpha rhythm gradually increases with age. Findings show a decrease in temporal windows of alpha wave postnatally through early childhood (Cellier et al., 2021; Freschl et al., 2022). This suggests rapid changes in visual information processing that implies residual firing rates resulting from overexcitability and a level of behavioral disorganization in early gestation, both reducing gradually with age from mid-gestation onwards.

Opening and closing of the eyes changes the power of the alpha wave, and thus it reflects transitory alert behavioral states which accompany early executive functions, as shown in studies of preterm infants born very early before term (Als et al., 2004, 2012). Notably, alpha peaks were shown to be associated directly with changes in white matter and healthy volume of cortical regions as well as with network connectivity, myelination, and axonal growth, especially in the thalamocortical pathways. Findings show that coordinated signaling between the young thalamus and early premature posterior cortical regions appears with improved generation of alpha waves (Martini et al., 2021). This gradual growing coordination of firing rate during fetal behaviors after mid-gestation was also implied for the entire prefrontal-thalamic-hypothalamic-hippocampal-corticospinal tract network (Elisevich et al., 2025; López-Bendito, 2018; Sherman, 2012; Stein et al., 2000). Sleep state discrimination and wakefulness patterns mature through the inhibitory function of the postsynaptic GABAergic and cholinergic neuro-activity which interact to regulate the levels of excitatory norepinephrine and glutamate (Goldstein Ferber et al., 2022). This occurs upon the changes in the GABAergic signaling turning at mid-gestation from excitatory to its traditional inhibitory role, leaving the excitatory role mainly to glutamate, as widely accepted (Herlenius & Lagercrantz, 2004).

Thus, the development of the continuous EEG patterns and especially first alpha and then beta wave emergence from the theta-delta wave, promoting increasing low-voltage fast activity of wakefulness and cortical functions, represents the balance between excitatory and inhibitory terminals. This is not reached before mid-gestation, and it presents a turning point from overexcitability during the embryonic phases into the onset of organismal excitation-inhibition balance. The change in the EEG patterns at this turning point marks a volumetric maturation of the cortical-thalamic-hypothalamic-hippocampal-corticospinal tract network connectivity and reduced differentiation, while cell specification and coordinated firing rate are increased. This in turn is shown in transitory eye opening and alert state accompanying more organized fetal behaviors that can be related to an outside input, and thus it can be termed as the origination of very early executive functions.

To afford further differentiation and proliferation within the future network underlying the elicitation of executive functions, timewise clocks of progenitor expression that are crucial for further volume increase of the substrates employ a unique molecular balance between expressing and expression-stopping molecules (Sladitschek & Neveu, 2019). This functional-structural uniqueness derived from its molecular sensitivity to the time during gestation, working as clocks, is herewith further detailed relating specifically to each of the developing regions in the network responsible for early executive functions, the prefrontal-thalamic-hypothalamic-hippocampal-corticospinal tract. For each region, the time-specific area-specific combinatorial impact of its molecular basis of development and maturation along gestation is described.

The prefrontal cortex, the main region responsible for originating executive functions, expands in different rates and times during embryonic phases independent of growth rate and volumes of other regions. The specific growth factors within the developing prefrontal cortex are active before they receive messages from afferent axonal inputs from lower regions. Intersecting gradients of molecules are expressed across the embryonic cerebral wall that later attract afferent fibers to their appropriate positioning and before those fibers interact with neighboring cell sets in the developing prefrontal cortex. The onset of prefrontal development is initiated by an elevation in the number of symmetrical divisions of neural stem cells in the ventricular zone before the start of neurogenesis. This is followed by the formation of the subventricular intermediate subplate and cortical plate zones (Rakic, 1988). This affords massive increase in the cortical surface area. Differential rates of growth in the different prefrontal growing systems (Kolk & Rakic, 2022) occur in parallel sequences with lower regions in the future prefrontal-thalamic-hypothalamic-hippocampal-corticospinal tract network. This suggests that the cortical and subcortical regions develop early structural volumes towards each other, and their mutual embrace elicits the components of the alert state and more complex behavioral phenomena later in gestation.

In detail, a nonsymmetrical and nonlinear progress of prefrontal development has been recorded recently (Kolk & Rakic, 2022). By the time that apical and glial progenitors within all prefrontal zones start dividing asymmetrically, the number of neurons increases and peaks at 13–16 weeks of gestation in humans. Along this embryonic progress, cell-autonomous local zone-related events continue to increase, supported by glial cells. Importantly, this progress continues to the late stages of gestation according to recent studies that showed the role of a new class of progenitors. These new classes of progenitors are actively dividing as non-epithelial progenitors that can maintain contact with the pia but not the ventricle with radial morphology and expression of nuclear and cytoplasmic markers, termed as glial-like and found in the outer subventricular zone. These progenitors are involved in volume growth and specification of the superior frontal, inferior frontal, and precentral regions at 25–26 weeks of gestation (Hansen et al., 2010; Malik et al., 2013). This finding stays in contrary to other claims that progenitor impacts are silenced after mid-gestation, favoriting environmental effects (Vasung et al., 2019). Thus, striking as it may be, the genetic and cellular building blocks of executive functions develop in timewise rhythms along the entire period of gestation.

Specifically, according to recent findings, horizontal positioning of neurons (surface size) is dependent on the positions of precursor cells, while vertical growth (thickness) is dependent on the time of their origin (Kolk & Rakic, 2022). During these clock-driven prefrontal developmental stages, which function along the entire period of gestation, stereotyped behaviors dependent on partially available and not coordinated neural bursts are produced earlier in the embryo, while from mid-gestation onward these molecular impacts permit fetal behaviors to become gradually experienced-based too.

The thalamus, the main gateway for somatosensory processing as well as being the hub for information processing and attention, is also formed in a spatiotemporal unique manner (Kostović, 2024). In this central region, striking findings show that progenitor expression timing determines the identity of the cells in the thalamic area. Shh-responsive cTh.Pro1 progenitors give rise first to glutamatergic neurons within sensory, motor, and anterior nuclei for the development of the ventrolateral areas. Then, Shh-responsive rTh.Pro progenitors elicit GABAergic neurons in the thalamic nuclei that form the dorsomedial positions of the thalamus. Shh-independent Cth.Pro2 progenitors induce the birth of neurons within the nuclei in a single-cell outside-in manner along the time of the thalamic areas' generation (Kostović, 2024; Govek et al., 2022). This timewise process is readily distinguished by mid-gestation close to the time of the thalamic fibers' penetration into the cortical plate (Kostović, 2024), the onset of wide eye opening in the behavioral repertoire of the fetus, and its more organized stimulus-responsive nonrandom behaviors (Donovan et al., 2020; Fagard et al., 2018).

The hypothalamic areas, the main regions regulating residual excitatory conditions, also develop from timely expression of specific progenitors, especially Shh/Fgf10, which elicit the expression of other progenitors that affect volumetric factors and area specifications in this region. Shh are expressed early in gestation within the embryonic phase and contribute at these early times to the morphology and formation of the neural tube. Their expression determines cell fate in the neural tube and floor plate. Their special impact on the development and growth of the hypothalamic areas is presented by time-specific expression concentrations with timewise trajectories along the entire period of gestation up to hypothalamic inter-regional connectivity with other cerebral regions responsible for fetal executive functions in healthy gestation (Fu et al., 2019). Thus, the progenitors involved in hypothalamic development express unique characteristics of sensitivity to times working as clocks by expressing different concentrations at different times along gestation for area identity and inter-regional connectivity.

As the hippocampus is the main region for storing, processing, and projecting information on memory and learning, its development is considered here as crucial for the elicitation of executive functions (Griffin, 2015). The important findings on progenitor clocks show their contribution to neural birth in the hippocampus before mid-gestation and after as well as their impact on the hippocampal volume and levels of functional neural connectivity. At this point, the state of this art shows that early progenitors affect specific early-born neurons in specific areas of the hippocampus, while other progenitors are active later for other specific regions in the hippocampus with later-born neurons and later maturity. The changes of fetal behaviors from stereotyped cyclical and random to more organized and stimulus-related ones have been reported in recent years (Einspieler et al., 2021; Fagard et al., 2018). The period between 22 and 32 weeks of gestation can thus be partly related to the differential progress of progenitors responsible for hippocampal development and their nonlinear elicitation of gene expression levels (Yang et al., 2014). Specifically, the expression of the endogenous proteins Ki67 and proliferation cell nuclear antigen (PCNA) is associated with cell proliferation in this region. It shows a timely balance between gene expression and molecules that block this expression, a balance that provides the opportunity of expression restriction to specific areas at specific timings during gestation. A recent study (Yang et al., 2014) showed that Ki67-positive cells decrease in humans from 9 weeks of gestation to 24 weeks of gestation, while the mature neurons in the hippocampus appeared in cornu ammonis as early as 9 weeks of gestation and were found to increase and be active at 22 weeks of gestation, the time when a striking increase of the number of mature neurons was found. Before birth, more progenitor cells were found in the dentate gyrus than in the cornu ammonis, showing that different balanced molecular levels are required as gestation progresses and that these balances are area-specific and time-specific.

The corticospinal tract (CST) is a crucial region for the production of motor behaviors, developing from mid-gestation until 2 years of age postnatally (Braga et al., 2015). The CST is also the largest connection between the brainstem (which matures at mid-gestation) and the cortex (which develops later in its six-layered form between 24 and 32 weeks of gestation). Recent findings showed that the CST is involved in early executive functions (Goldstein Ferber et al., 2022).

Other findings showed that specific gene subtype expression is time-dependent and area-specific. Specific subtypes are expressed in a given cortical area, while other subtypes are parallelly excluded from expression in the same area in healthy gestation. For example, CTIP2 is a prerequisite for healthy gestation, and its absence during the birth of corticospinal motor neurons in unhealthy gestation results in motor and degenerative diseases. However, in healthy gestation, it is excluded from interhemispheric callosal neurons, which project to the midline (Fame et al., 2011), although co-localized in the same cortical layer with other subtypes which project in a downstream manner (Arlotta et al., 2005). In addition to being area-specific, according to recent reports, the corticospinal tract genetic clocks of expression for the birth of motor neurons include genes expressing in a timewise manner along gestation: (1) in early development, in the embryonic phase; (2) in the intermediate period of gestation during the end of the first semester; and (3) during late gestation, after mid-gestation. This timewise expression controls early as well as late development of motor neurons (Arlotta et al., 2005).

The required progenitors' expression and excluded expression were found to work in concert at specific timings in a combinatorial manner. This means that findings show a specific combination of expressions which are also region-specific and time-specific. These clocks of gene expression produce the corticospinal motor neurons involved in the origination of fetal behavior. This combinatorial manner suggests that only specific combinations at specific timings along gestation can further continue towards motor neuron production and aligned behaviors. The specific required combinations are not restricted to the embryonic phase, and their impact at specific timings is dependent on feedback loops from produced behaviors as gestation develops and especially after mid-gestation onwards.

As a result of these area-specific, time-specific molecular events, intra- and inter-regional polarization depolarization events are relatively coordinated later in gestation, bringing about the behavioral changes from embryonic to fetal phases. Thus, the turning point of prefrontal-thalamic-hypothalamic-hippocampal-corticospinal tract networking development (Arlotta et al., 2005; Ferber et al., 2022; Goldstein Ferber et al., 2022; Griffin, 2015; Klein et al., 2020; Kolk & Rakic, 2022; Yang et al., 2014) towards the production of more organized stimulus-responsive fetal execution functions during short and transitory awake states is apparent at the developmental stage of wide fetal eye opening, following visuomotor maturation at 28 weeks of gestation. Notably, the timewise developmental coordination of action potentials within the prefrontal-thalamic-hypothalamic-hippocampal-corticospinal tract network from mid-gestation onwards gradually eliminates propagation delays and the duration of residual excitation as healthy gestation progresses.

In sum, rhythms of time – and area specific progenitors induce the timely elicitation of the earliest executive functions. Understanding the timewise origin of early executive functions' occurrence provides the appreciation of the transition between early and late gestational clockwise order, taking into account the time window of mid-gestation. While the molecular basis, affording the occurrence of short alert states and early fetal executive functions, encompasses a time-sensitive mode of expression, mid-gestation is the timing when genetic and cellular clocks are beginning to be shaped by experience, suggesting that at this time window any remaining differentiation starts to overlap with pruning processes. At this turning time point in gestation, coordinated connectivity in the prefrontal-thalamic-hypothalamic-hippocampal-corticospinal tract pathways occurs, as shown in fetal brain wave production, affording feedback and feedforward gene-environment loops.

It is hypothesized that this time-sensitive transition around mid-gestation is afforded by changes in the earlier combinatorial balance between molecules which promote and those which restrict expression. It is suggested that the times of these developing combinations are governed by clock genes. It is further suggested that this later developing balance gradually changes on the genetic-driven level to respond to environmental inputs too, at different times, similarly to the observed photic reactivity of clock genes. The capacity of clock genes to elicit timely genome-wide molecular sensitivity to the environment by their early responsivity to it, is based on findings that in addition to their sensitivity to the environment, they generate the transcriptomic architecture of other genes. (A detailed explanation is provided in chapter 12, the concluding chapter, including an integrating model) (Peng et al., 2019). Thus, clock genes persist to move the fetal organism forward in the development of healthy gestation by their combined sensitivity to molecular balances (which then create neural structures and connectivity) and to environmental changes. As such, they design the times of elicited molecular combinations, gestational-age appropriate. From the turning point of mid-gestation onwards, these clocks interact with the environment too, in a manner that any fetal acquisition of an executive function cannot be predicted by either the clocks or the environmental stimuli only.

References

Als, H., Duffy, F. H., McAnulty, G. B., Rivkin, M. J., Vajapeyam, S., Mulkern, R. V., Warfield, S. K., Huppi, P. S., Butler, S. C., Conneman, N., Fischer, C., & Eichenwald, E. C. (2004). Early experience alters brain function and structure. *Pediatrics, 113*(4), 846–857. https://doi.org/10.1542/peds.113.4.846

Als, H., Duffy, F. H., McAnulty, G., Butler, S. C., Lightbody, L., Kosta, S., Weisenfeld, N. I., Robertson, R., Parad, R. B., Ringer, S. A., Blickman, J. G., Zurakowski, D., & Warfield, S. K. (2012). NIDCAP improves brain function and structure in preterm infants with severe intrauterine growth restriction. *Journal of Perinatology, 32*(10), 797–803. https://doi.org/10.1038/jp.2011.201

Arlotta, P., Molyneaux, B. J., Chen, J., Inoue, J., Kominami, R., & Macklis, J. D. (2005). Neuronal subtype-specific genes that control corticospinal motor neuron development in vivo. *Neuron, 45*(2), 207–221. https://doi.org/10.1016/j.neuron.2004.12.036

Braga, R. M., Roze, E., Ball, G., Merchant, N., Tusor, N., Arichi, T., Edwards, D., Rueckert, D., & Counsell, S. J. (2015). Development of the corticospinal and Callosal tracts from extremely premature birth up to 2 years of age. *PLoS One, 10*(5), e0125681. https://doi.org/10.1371/journal.pone.0125681

Britton, J. W., Frey, L. C., Hopp, J. L., Korb, P., Koubeissi, M. Z., Lievens, W. E., Pestana-Knight, E. M., & St. Louis, E. K. (2016). In E. K. St. Louis & L. C. Frey (Eds.), *Electroencephalography (EEG): An introductory text and atlas of Normal and abnormal findings in adults, children, and Infants*. American Epilepsy Society. http://www.ncbi.nlm.nih.gov/books/NBK390354/

Cellier, D., Riddle, J., Petersen, I., & Hwang, K. (2021). The development of theta and alpha neural oscillations from ages 3 to 24 years. *Developmental Cognitive Neuroscience, 50*, 100969. https://doi.org/10.1016/j.dcn.2021.100969

Donovan, T., Dunn, K., Penman, A., Young, R. J., & Reid, V. M. (2020). Fetal eye movements in response to a visual stimulus. *Brain and Behavior, 10*(8), e01676. https://doi.org/10.1002/brb3.1676

Einspieler, C., Prayer, D., & Marschik, P. B. (2021). Fetal movements: The origin of human behaviour. *Developmental Medicine & Child Neurology, 63*(10), 1142–1148. https://doi.org/10.1111/dmcn.14918

Elisevich, L., Ascenzi, B. M., & Elisevich, K. (2025). Chapter 7 – The diencephalon: Introduction and Epithalamus. In B. M. Ascenzi (Ed.), *From anatomy to function of the central nervous system* (pp. 229–254). Academic Press. https://doi.org/10.1016/B978-0-12-822404-5.00002-4

Fagard, J., Esseily, R., Jacquey, L., O'Regan, K., & Somogyi, E. (2018). Fetal origin of sensorimotor behavior. *Frontiers in Neurorobotics, 12*. https://doi.org/10.3389/fnbot.2018.00023

Fame, R. M., MacDonald, J. L., & Macklis, J. D. (2011). Development, specification, and diversity of callosal projection neurons. *Trends in Neurosciences, 34*(1), 41–50. https://doi.org/10.1016/j.tins.2010.10.002

Ferber, S. G., Als, H., McAnulty, G., Klinger, G., & Weller, A. (2022). Multi-level hypothalamic neuromodulation of self-regulation and cognition in preterm infants: Towards a control systems model. *Comprehensive Psychoneuroendocrinology, 9*, 100109. https://doi.org/10.1016/j.cpnec.2021.100109

Ferber, S. G., Geva, R., & Weller, A. (2023). When the mind comes to live inside the body: The ontogeny of the perceptual control clock. *Current Neuropharmacology, 21*(1), 13–21. https://doi.org/10.2174/1570159X20666220411095508

Freschl, J., Azizi, L. A., Balboa, L., Kaldy, Z., & Blaser, E. (2022). The development of peak alpha frequency from infancy to adolescence and its role in visual temporal processing: A meta-analysis. *Developmental Cognitive Neuroscience, 57*, 101146. https://doi.org/10.1016/j.dcn.2022.101146

Fu, T., Pearson, C., Towers, M., & Placzek, M. (2019). Development of the basal hypothalamus through anisotropic growth. *Journal of Neuroendocrinology, 31*(5), e12727. https://doi.org/10.1111/jne.12727

Goldstein Ferber, S., Weller, A., Ben-Shachar, M., Klinger, G., & Geva, R. (2022). Development of the ontogenetic self-regulation clock. *International Journal of Molecular Sciences, 23*(2) Article 2. https://doi.org/10.3390/ijms23020993

Govek, K. W., Chen, S., Sgourdou, P., Yao, Y., Woodhouse, S., Chen, T., ... & Camara, P. G. (2022). Developmental trajectories of thalamic progenitors revealed by single-cell transcriptome profiling and Shh perturbation. *Cell Reports, 41*(10).

Griffin, A. L. (2015). Role of the thalamic nucleus reuniens in mediating interactions between the hippocampus and medial prefrontal cortex during spatial working memory. *Frontiers in Systems Neuroscience, 9*. https://doi.org/10.3389/fnsys.2015.00029

Hansen, D. V., Lui, J. H., Parker, P. R. L., & Kriegstein, A. R. (2010). Neurogenic radial glia in the outer subventricular zone of human neocortex. *Nature, 464*(7288), 554–561. https://doi.org/10.1038/nature08845

Herlenius, E., & Lagercrantz, H. (2004). Development of neurotransmitter systems during critical periods. *Experimental Neurology, 190*, 8–21. https://doi.org/10.1016/j.expneurol.2004.03.027

Hopkins, B., & Johnson, S. P. (Eds.). (2005). *Prenatal development of postnatal functions*. Praeger.

Jabbari, S. (2021). Source separation from single-channel abdominal phonocardiographic signals based on independent component analysis. *Biomedical Engineering Letters, 11*(1), 55–67. https://doi.org/10.1007/s13534-021-00182-z

Klein, B., Mrowetz, H., Kreutzer, C., Rotheneichner, P., Zaunmair, P., Lange, S., Coras, R., Couillard-Despres, S., Rivera, F. J., & Aigner, L. (2020). DCX+ neuronal progenitors contribute to new oligodendrocytes during remyelination in the hippocampus. *Scientific Reports, 10*(1), 20095. https://doi.org/10.1038/s41598-020-77115-w

Kolk, S. M., & Rakic, P. (2022). Development of prefrontal cortex. *Neuropsychopharmacology, 47*(1), 41–57. https://doi.org/10.1038/s41386-021-01137-9

Kostović, I. (2024). Development of the basic architecture of neocortical circuitry in the human fetus as revealed by the coupling spatiotemporal pattern of synaptogenesis along with microstructure and macroscale in vivo MR imaging. *Brain Structure and Function, 229*(9), 2339–2367. https://doi.org/10.1007/s00429-024-02838-9

López-Bendito, G. (2018). Development of the Thalamocortical interactions: Past, present and future. *Neuroscience, 385*, 67–74. https://doi.org/10.1016/j.neuroscience.2018.06.020

Malik, S., Vinukonda, G., Vose, L. R., Diamond, D., Bhimavarapu, B. B. R., Hu, F., Zia, M. T., Hevner, R., Zecevic, N., & Ballabh, P. (2013). Neurogenesis continues in the third trimester of pregnancy and is suppressed by premature birth. *Journal of Neuroscience, 33*(2), 411–423. https://doi.org/10.1523/JNEUROSCI.4445-12.2013

Martini, F. J., Guillamón-Vivancos, T., Moreno-Juan, V., Valdeolmillos, M., & López-Bendito, G. (2021). Spontaneous activity in developing thalamic and cortical sensory networks. *Neuron, 109*(16), 2519–2534. https://doi.org/10.1016/j.neuron.2021.06.026

Mizrahi, E. M., & Hrachovy, R. A. (2015). *Atlas of neonatal electroencephalography*. Springer.

Myers, M. M., Fifer, W. P., Grose-Fifer, J., Sahni, R., Stark, R. I., & Schulze, K. F. (1997). A novel quantitative measure of Tracé-alternant EEG activity and its association with sleep states of preterm infants. *Developmental Psychobiology, 31*(3), 167–174. https://doi.org/10.1002/(SICI)1098-2302(199711)31:3<167::AID-DEV1>3.0.CO;2-Q

Myowa-Yamakoshi, M., & Takeshita, H. (2006). Do human fetuses anticipate self-oriented actions? A study by four-dimensional (4D) ultrasonography. *Infancy, 10*(3), 289–301. https://doi.org/10.1207/s15327078in1003_5

Peng, G., Suo, S., Cui, G., Yu, F., Wang, R., Chen, J., ... & Jing, N. (2019). Molecular architecture of lineage allocation and tissue organization in early mouse embryo. *Nature, 572*(7770), 528–532.

Rakic, P. (1988). Specification of cerebral cortical areas. *Science, 241*(4862), 170–176. https://doi.org/10.1126/science.3291116

Reid, V. M., Dunn, K., Young, R. J., Amu, J., Donovan, T., & Reissland, N. (2017). The human fetus preferentially engages with face-like visual stimuli. *Current Biology, 27*(12), 1825–1828.e3. https://doi.org/10.1016/j.cub.2017.05.044

Reissland, N., Francis, B., Aydin, E., Mason, J., & Schaal, B. (2014). The development of anticipation in the fetus: A longitudinal account of human fetal mouth movements in reaction to and anticipation of touch. *Developmental Psychobiology, 56*(5), 955–963. https://doi.org/10.1002/dev.21172

Sherman, S. M. (2012). Thalamocortical interactions. *Current Opinion in Neurobiology, 22*(4), 575–579. https://doi.org/10.1016/j.conb.2012.03.005

Sladitschek, H. L., & Neveu, P. A. (2019). A gene regulatory network controls the balance between mesendoderm and ectoderm at pluripotency exit. *Molecular Systems Biology, 15*(12), e9043. https://doi.org/10.15252/msb.20199043

Spoto, G., Saia, M. C., Amore, G., Gitto, E., Loddo, G., Mainieri, G., Nicotera, A. G., & Di Rosa, G. (2021). Neonatal seizures: An overview of genetic causes and treatment options. *Brain Sciences, 11*(10), Article 10. https://doi.org/10.3390/brainsci11101295

Stein, T., Moritz, C., Quigley, M., Cordes, D., Haughton, V., & Meyerand, E. (2000). Functional connectivity in the thalamus and hippocampus studied with functional MR imaging. *American Journal of Neuroradiology, 21*(8), 1397–1401.

Thaler, I., Boldes, R., & Timor-Tritsch, I. (2000). Real-time spectral analysis of the fetal EEG: A new approach to monitoring sleep states and fetal condition during labor. *Pediatric Research, 48*(3), 340–345. https://doi.org/10.1203/00006450-200009000-00013

Vasung, L., Abaci Turk, E., Ferradal, S. L., Sutin, J., Stout, J. N., Ahtam, B., Lin, P.-Y., & Grant, P. E. (2019). Exploring early human brain development with structural and physiological neuroimaging. *NeuroImage, 187*, 226–254. https://doi.org/10.1016/j.neuroimage.2018.07.041

Yang, P., Zhang, J., Shi, H., Zhang, J., Xu, X., Xiao, X., & Liu, Y. (2014). Developmental profile of neurogenesis in prenatal human hippocampus: An immunohistochemical study. *International Journal of Developmental Neuroscience, 38*, 1–9. https://doi.org/10.1016/j.ijdevneu.2014.06.015

Chapter 11
The Clocks of Emotions: Rhythms of Molecular Sensitivity to Experience-Shaped Synaptic Loops Breeds Emotions

Abstract The novel techniques of fetal MRI, 3D and 4D ultrasound recordings have now made it possible to outline the development of neurites emerging parallelly in preparatory cortices and subcortical regions toward each other, as well as the trajectories of progressing behavioral organization originated in earlier random behaviors. Thus, it is hard to conclude that the first affective experience in healthy early life is narrowed to stress or pain reactivity or exposure to stress only as thought before. It is rather more inclusive to conclude that during healthy gestation the clocks of emotion are dependent on intricate but regulated in parallel sequences of regional development, axonal outgrowth between these regions, aligned dendritic arborization, and endocrine development, counter to the hypothesis of a distinct linear bottom-up timeline of brain development. Recent evidence from 3D and 4D ultrasound recordings of fetal behaviors showed that from mid-gestation onwards, repeated self-touching motor behaviors, especially toward highly innervated regions such as the face and the mouth, are related to a sensory trigger. Taken together, the complex infrastructure basis for the appearance of emotions is not all about axonal growth and not all about bottom-up development only. It is about controls and balances during healthy gestation for the clocks of emotion to develop. It is shown in this chapter that clock genes design the timing of DNA molecular sensitivity to environmental stimuli at mid-gestation. Pruning and experienced-shaped molecular events are available by the time molecular sensitivity to the environment starts. The clocks of emotion are part of the fetal experience which contributes to behavioral feedback loops for pruning purposes. From an evolutionary perspective, the clocks of emotion prepare the fetus to human interaction for survival and growth outside the womb.

What is the first emotion in life? When does it start to be experienced in humans? And what are its biological pathways of development? To answer these questions, this chapter will describe the clockwise development of early emotional reactivity as arising from the progress of fiber arborization, increasing connections between regions such as the HPA axis and the corticolimbic areas, including their connectivity with other developing cortical areas.

S. Goldstein Ferber, *Very Early Clocks*,
https://doi.org/10.1007/978-3-031-89603-3_11

Before asking the question, what is the timing of the first feeling or affective experience during gestation or whether it appears postnatally, there is a preceding question asking whether consciousness, or just a partial state on the dynamic flow of the human consciousness spectrum, is required for very early affective sensitivity. The answer to this question may have implications for medical procedures and indications to treat very-early-born preterm infants. This chapter presents affective development of the embryonic and fetal motor spontaneous and reactive behaviors as they relate to the clockwise trajectories of regional development, axonal outgrowth between these regions, and aligned dendritic arborization as well as early endocrine development.

Earlier considerations related consciousness, defined by a narrow definition of awareness to the body, and the emergence of a body-self, to the borderline of viability around mid-gestation (Lagercrantz & Changeux, 2009). To support this view, studies of preterm infants showed that pain is experienced by this timing during gestation including observations of learned avoidance behavior shown in preterm infants responding to environmental conditions of previous painful experiences (Bauer, 2006). Recent evidence from 3D and 4D ultrasound recordings of fetal behaviors showed that from mid-gestation onwards, repeated self-touching motor behaviors, especially towards highly innervated regions such as the face and the mouth, are related to a sensory trigger. These data reflect the onset of behavioral integration by the trigger-output link and a more mature behavioral organization compared to random behaviors of the embryo. This period of time in gestation has been termed earlier as a "fetal search for a body-map" (Fagard et al., 2018).

Most of the research on the elicitation of affective experiences in early life concentrated on early stress, reaching the fetus either by maternal exposure or placental insufficiency (Guadagno et al., 2021). The new techniques of fetal MRI, 3D and 4D ultrasound recordings have now made it possible to outline the development of neurites emerging parallelly in preparatory cortices and subcortical regions towards each other as well as the trajectories of progressed behavioral organization originated in earlier random embryonic behaviors. Thus, it is hard to conclude that the first affective experience in healthy early life is narrowed to stress or pain reactivity. It is rather more inclusive to conclude that during healthy gestation the clocks of emotion are dependent on intricate but regulated parallel sequences of regional development, axonal outgrowth between these regions, aligned dendritic arborization, involved neurotransmitter signaling, and endocrine development, counter to the hypothesis of a distinct linear bottom-up timeline of brain development.

Thus, it is suggested that the early affective experiences accumulate to elaborate the binary spectrum of pain vs. reward experiences, from bad-like vs. good-like feelings during early stages of gestation to more robust sensory-motor links, growing from inner triggers (Lagercrantz & Changeux, 2009) to include environmentally triggered emotions too.

Behavioral differentiation as a developmental challenge for preterm infants has been earlier shown (Ferber et al., 2011). While it is widely accepted that emotions accompany behavioral output in mammals, more complex and diverse feelings may arise in the fetal organism with a given behavioral output, while embryonic behavior is apparent before fetal emotionality is aligned with motor movements. This means that a given behavioral output can be conditioned to more than one fixed type of feeling when repeated, a gradual process termed here as *developmental affective differentiation*. Earlier data show that such behavioral-affective experiences, although based on sensorimotor maturation, may even be stored for a short time during fetal life, as exemplified by the phenomenon of habituation. One well-accepted example is that at 22–23 weeks of gestation, if a human fetus is exposed to a repetitive sound, it reacts by movements but habituates after repeated exposures to the same sound (Leader et al., 2009). Thus, the feeling that accompanies the motor reactions to the first stimuli is different from the feeling in the habituated phase. Since it is stored for a short time only, the next exposure may arise different motor behaviors and different accompanying feelings. This suggests the accumulating diversity of emotional reactions to the same stimulus, a gradual intrauterine process that peaks near term, preparing the fetus to the extrauterine multi-stimuli environment. This is an integral part of accomplishing the next peaks postnatally.

Thus, the process of accumulating affective experiences occurs through affective differentiation of the sensory-to-motor ratio. These growing compounded emotions may be typical within an accumulating spectrum ranging between bad-like and good-like feelings. This range of accumulating emotions with feedback mechanisms aiming towards more and more behavioral organization, although originating from disorganized behavioral output during embryonic life, may be possible only if the neural excitation-inhibition balance is kept appropriate for gestational age and related metabolic demands. Otherwise, elevated oxidative stress or reduced timely cell polarization and neural signaling may result in periods of either conditions of excitotoxicity or extreme reduction in the frequency of behaviors, with tissue, molecular, and metabolic insults. Experiencing the extremes of excitation-inhibition balance determines the declines in the healthy progress of affectivity development and the clocks of emotion, preventing accumulation and diversity in fetal reactive feelings.

The establishment of the clocks of emotions is a prolonged process which occurs between 8 and 34 weeks of gestation. During very early stages of healthy gestation, emotional reactions occur as blitzes that accompany random behaviors prior to mid-gestation. Around mid-gestation these emotional reactions become prolonged (in seconds) towards more and more differentiated types of emotions when they accompany more complex and repeated behaviors along with the functional complexity of mutual cortical-limbic projection loops being established. This phase is followed by a period of increased behavioral complexity and emotional differentiation as a preparation to the out-uterine environment starting from 34 weeks of gestation onwards. The complex behaviors and the elimination of prior random behaviors found from mid-gestation onwards result from a cortical-subcortical mutual axonal growth and dendritic arborization and depend on the balance between molecular attracting and restraining factors.

The building blocks of the clocks of emotions, as derived from histological and fetal MRI studies, start early in gestation, around 10 weeks of gestation, and peak around mid-gestation. This is due to structural and functional developments between the developing six-layered cortex core and the surrounding primitive laminar architecture of the limbic subcortical regions regulating emotional reactivity, especially the septum which projects to the hypothalamus, hippocampus, amygdala, reticular formation, and olfactory cortical regions. Through these basal telencephalic, diencephalic, mesencephalic, and rhombencephalic control systems, the limbic system becomes the first source for regulating expectancies in the cortical hierarchy (Tucker & Luu, 2021) and the developing large association neurite (Han & Šestan, 2013).

The limbic bundles develop at a very early stage. The most prominent limbic bundle is the fornix, and it connects with the septum. By 10 weeks of gestation, major afferent fibers invade the cortico-striatal crossroad. Between 15 and 18 weeks of gestation, the periventricular junctions elaborate, while penetration of the thalamic fibers into the cortical plate occurs after 24–26 weeks following a period of "waiting" in the subcortical plate. Findings show that in the early gestational period (11–17 weeks of gestation), several limbic bundles were identified, including the fornix, followed by the accumulation of afferent thalamocortical axons in the intermediate and cortical subplate zone (17–25 weeks of gestation), which in turn is followed by the structural onset (26–34 weeks of gestation) and final formation of associative bundles (35–40 weeks of gestation) (Vasung et al., 2010). Although the above may seem as just a bottom-up development, recent imaging data show that between 29 and 31 weeks, the thalamocortical connectome is observable, while cortical-cortical network connections are detectable too (Taymourtash et al., 2023). Previous data show that top-down processes occur in parallel at the same timings, even earlier in gestation, implying that the clocks of emotions depend on timely axonal growth originated in early developing cortical projection neurons targeting the internal capsule and the thalamus (Richards et al., 1997). Before cortical neurogenesis starts, and in order to increase their availability and number, neuroepithelial progenitor cells are processed through a series of symmetric divisions, and then they differentiate into early radial glia cells. Neurons generated from these cells then migrate along radial glia fibers toward the cortical surface (Vasung et al., 2010).

Earlier findings show that cortical axonal growth which targets the internal capsule is controlled chemically for pathfinding by chemoattractants of molecules such as inhibitory netrin-1 and excitatory Fezf2. Emerging radial glial cells act as multipotent progenitors that sequentially generate all subtypes of these early cortical projection neurons. Findings showed that cortical axonal outgrowth was directed towards the internal capsule, when at day 15 of gestation explants from rat neocortex and intermediate targets were cultured but suppressed when cocultured with medial wall of the dorsal telencephalon tissue. This suggests a balance between factors which increase growth and factors with a "stop signal" to govern the rate of growth. Thus, the internal capsule releases chemoattractants for cortical axons, whereas the medial wall of the dorsal telencephalon releases chemosuppressants (Richards et al., 1997). Additionally, as excitatory factors dominate very early development, the complexity of early cortical downstream projections is exemplified by striking variations found in the origin and numbers of inhibitory interneurons. It was shown that timing of functional availability of these early-born neurons depends on telencephalic regional origin while, cell specification depends on voltage gating, resulting from the functions of ion channels and ATPase pumps (Fishell & Rudy, 2011).

The different types of cortical projection neurons function to target lower situated regions, such as the limbic system, through periventricular junctions, the internal capsule, and the thalamus (Carlos & O'Leary, 1992; Richards et al., 1997). Thus, lower and higher regions develop connectivity in parallel towards the other, with peaks in structural thickness and in the achievement of large complex network connectivity at the same time of mid-gestation. This is a time in gestation which is associated with more complex behavioral activity. The rate of axonal growth is controlled by the balance between growth promoting factors and other suppressing ones, originating in progenitors as early as 8–10 weeks of gestation. Therefore, the complex infrastructure basis for the appearance of emotions is not all about axonal growth and not all about bottom-up development only. It is about controls and balances during healthy gestation for the clocks of emotion to develop.

Cortical neurons have complex branches of dendritic arborization essential for their function. Dendrite formation occurs after a period of excessive, random, and spontaneous firing, especially through thalamocortical pathways. This occurs before radial migration is completed and following a prolonged time of proliferation. The shift between migration and arborization depends on the removal of inhibitory Sox11 gene that binds to specific regions in the DNA, coordinating other proteins, to elicit expression of other genes responsible for the birth of neuron cells (Hoshiba et al., 2016). Thus, dendritic arborization is elicited by intrinsically orchestrated genetic factors. However, later in gestation, this process is affected by activity-dependent structural changes (Arikkath, 2012; Menon & Gupton, 2018; Urbanska et al., 2008) to form the mature function of information input and transmission as well as synapse quantity and synapse selection. It also increases later dependence on cell specifications for maturation of corticolimbic connectivity loops (Goikolea-Vives & Stolp, 2021; Moreno-Juan et al., 2017; Pan & Monje, 2020).

The earliest cortical circuits in humans are formed in the preparatory cortical plate by gestational week 5 (Allendoerfer & Shatz, 1994; Supèr et al., 1998; Tau & Peterson, 2010; Wood et al., 1992). Neurons within this preparatory organelle create provisional unstable synaptic connections with close targeted cells, waiting for migrating neurons to arrive and create more stable networks as the cortical laminate develops. This process is detectable in the first neurons which project outside the cortex. The progress of migration is followed by dendritic arborization onset (Goikolea-Vives & Stolp, 2021). Phases of dendritic arborization are initiated by a slow elongation, continued by approach-avoidance pathfinding, guided by coordinated progenitors, to keep excitation-inhibition stable, leading to a stabilization phase postnatally, which also serves as the trigger for the pruning process onset (Menon & Gupton, 2018; Urbanska et al., 2008).

Scientific research showed that the first dendritic branching occurs between 16 and 26 weeks of gestation, and the arborization becomes more complex at 36 weeks (Lu et al., 2013; Walkley et al., 2000). An established dendritic tree is observable at term when cortical neurons can be detected with 30–55% of their maximum length (Becker et al., 1984). The basal dendrites establish their complexity earlier and do not show higher complexity postnatally while apical dendrites become complex later (Becker et al., 1984). As cortical connectivity along gestation appears in an inner-first outer-second manner, the formed synapses are diverse in function and specialization, with appearance first within the inner cortical layers (V, VI), and later within the outer ones (II, III) (Goikolea-Vives & Stolp, 2021; Huttenlocher & Dabholkar, 1997; Luu & Tucker, 2023).

The transition to activity-dependent dendritic arborization during late gestational periods is affected by Cajal-Retzius (CR) cells which are a transient cell population and, in the neocortex, secrete reelin to instruct the radial migration of projection neurons (Jossin, 2020). Transition to early experience-dependent control of neural activity occurs during the late fetal period. Recent scientific data show that dendritic arborization is not entirely a deterministic process nor a linear one and that postnatal restoration of dendritic complexity is possible. Dendritic branching requires post-mitotic suppression of the RhoGTPase Cdc42 catalyzed by the novel Cdc42-GAP NOMA-GAP. Earlier data showed that loss of NOMA-GAP leads to decreased neocortical volume, a vast reduced complexity in the dendritic tree in association with hyperactivation of Cdc42. Functioning of cofilin, an active agent regulating dendritic arborization during the late stages of gestation, depends on NOMA-GAP expression and its effect on inhibition of Cdc42. Findings showed in NOMA-GAP-deficient animals that active cofilin has the potential to reactivate postnatal dendritic complexity (Rosário et al., 2012).

Although it may seem that the creation of neural networks by axonal growth and dendritic arborization and even the transition to experience-dependent neural signaling is explained by these structural processes, this may be only a partial conclusion without consideration of the developing endocrine picture. Descriptions of hypothalamic regional development along gestation as shown in imaging studies (Koutcherov et al., 2003) provide findings on linear progress stages of regional subdivisions. However, recent data show that fetal endocrine developmental pathways develop in a U-shaped manner rather than in a linear trend. At early gestation, the presence of fetal steroidogenic enzymes, required for the synthesis of cortisol was detected, while during mid-gestation they were suppressed and elevated again near term. A U shape describes the time windows of vulnerability during gestation too, early and late gestation (Yong Ping et al., 2015). During early gestation, the embryo depends on buffering maternal and placental transfers with the protection of steroidogenic enzymes, and during late gestation it depends on elevated cortisol secretion while metabolic demands increase.

CRH was found in fetal hypothalamic tissue extracts at 12–13 weeks' gestation but not before (Ackland et al., 1986). As embryonic development progresses to 14–16 weeks of gestation, CRH-immunoreactive fibers can be detected in the median eminence (Ackland et al., 1986; Bresson et al., 1987). From mid-gestation, the fetal hypothalamus stimulates pituitary ACTH production. However, studies of preterm infants show that sequential release of hypothalamic CRH, ACTH, and cortisol appears only at late gestation. From early to mid-gestation, transfer of maternal cortisol to the fetus inhibits fetal ACTH production (Pepe & Albrecht, 1995), and reduction in fetal circulatory release of maternal cortisol upregulates it (Pepe &

Albrecht, 1995; Schoof et al., 2001), thus elevating fetal cortisol synthesis. This process can be evolutionarily understood as protecting fetal development, in conditions of limited availability of fetal ACTH (Rehman et al., 2007). Thus, up to late gestation, an interplay of endocrine products between the placenta and fetus is apparent. The placenta supports regulatory increase in fetal cortisol, while maternal cortisol suppresses fetal ACTH production to protect organ development until the endocrine sequence of CRH, ACTH, and cortisol matures in fetal hypothalamic circulation before term birth (Howland et al., 2017). Other findings show that the presence of such accurate endocrine interplay determines the prefrontal-amygdala connectivity and direction dominance, either towards increased sensitivity to stress or towards regulation of excitation-inhibition balance by progenitor actors (Marchisella et al., 2021).

These findings imply that the binary split between good-like feelings and bad-like feelings, that is, between stressful and rewarding conditions, may be elaborated only in the presence of such a developing fetal endocrine sequence. This endocrine sequence protects organogenesis and organs' signaling, including axonal growth and dendritic arborization, and as such keeps the adequacy of emotional clocks for the developing behavioral reactions. Thus, the healthy clocks of emotions are dependent on the maturation of the guardians of neural connectivity in the form of the protective endocrine picture, not merely on structural factors and their developing transmissions, neither on stress/no stress maternal exposures only, and with reversible corrections in adverse conditions.

This does not imply the dominance of the developing endocrine picture but rather suggests an orchestrated timing and time windows of vulnerability, as the structural development of axonal growth and dendritic arborization coincides with the hypothalamic endocrine progress, hypothalamic-limbic neural connectivity, and the fetal behavioral sets along gestation. During the second wave of neogenesis, around 9–10 weeks of gestation (van Eerdenburg & Rakic, 1994), fetal behavior emerges involving the onset of whole-body movements and an adaptive startle behavior followed by a general movement. This timing also coincides the start of the cortical subplate creation (Einspieler et al., 2021) and exemplifies the clocks of emotional onset by blitzes of rapid emotional reactions to a startle and a speedy reactivity consolidation phase by the production of general movement. The development of fetal affectivity from blitzes to relatively prolonged emotional reactions seems to occur not before 20 weeks of gestation when diurnal rhythms can be detected and there are observable periods of quiet (inhibitory) vs. active sleep (excitatory). This occurs with a progress in limbic-hypothalamic connectivity, while low fetal ACTH prevents the elevation in the protective cortisol, close to the time of the cortical subplate resolution and the penetration of thalamic fibers into the cortical plate. Around 30 weeks of gestation, along with a more matured endocrine picture and increase in the protective cortisol production (Ferber et al., 2022), the fetus shows repetition of complex movements, increase in facial expressions, and eye movements (Einspieler et al., 2021). This is the time when more intricate emotions accompany the relative complexity of behaviors preparing the fetus to term and birth as shown in studies of preterm infants (Ferber et al., 2022).

However, unlike traditional perspectives, simple linear structural growth and endocrine development is not implied here to explain healthy gestation but rather a regulatory perspective arguing that healthy gestation depends on the balance between excitatory and inhibitory progenitors and its molecular derivates (Marchisella et al., 2021). In line with this argument, a recent study showed accelerated growth in the fronto-limbic circuit of infants born before term (Lee et al., 2021). The striking evidence in this imaging study was the greater lateralization of fronto-limbic areas in preterm infants, without involvement of cerebral injury or earlier maternal exposure to stress, compared to term infants. Thus, rather than a linear constant progression in structural growth, these data shed light on the importance of timely excitatory progenitors signaling in cerebral regions involved in the development of affectivity, such as the fronto-limbic circuits, their impact on hypothalamic modulation of structural growth (Ferber et al., 2022), and for the healthy progression of fetal intricate range of emotional development. Thus, the timely development of the very early clocks of emotions evolves from time-dependent changes in fetal excitation-inhibition molecular balance, affecting structural development and it's signaling-related functions within fronto-hypothalamic-limbic feedback loops. This suggests an orchestrated basis with structure-function back-and-forth progression. Timing counts for the cerebrum structure-function mutual effects, at least as the sensitivity to stress does. Timing of very early clocks of emotion, elicited by fetal behavior, follows this balance. For diagnostic purposes, changes in progression in fetal behavioral sets and in the timing of their appearance imply changes in the development of emotionality.

Recent findings shed light on the timewise design of structural and functional developments described in this chapter. It was shown that clock gene homologs define the onset and offset of axonal growth and dendritic arborization (Noda et al., 2019; Suzuki et al., 2022). Clock genes are also closely related to the molecular basis of endocrine development and the on-off regulation of ACTH secretion (Nader et al., 2010). Clock genes' capacity to control time-sensitive molecular events is embedded in their early ultradian oscillations and their impact on genome-wide transcriptomic architecture along the embryonic phase, shifting to circadian rhythms following mid-gestation (Cox & Takahashi, 2019; Yagita, 2024). Thus, the timely processes presented in this chapter as the basis affording the intricate development of the clocks of emotions are controlled in an onset and offset manner by clock genes along the structural nonlinear trajectory of emotion-related brain regional development. It is concluded that early clock gene expression occurs in peripheral clocks, which are tissue-specific and dependent on cell identity too. At mid-gestation the chrono-control by clock genes becomes apparent in a more integrated whole-region arousal or decline pattern. This occurs following structural development of the suprachiasmatic nucleus at mid-gestation coinciding with the onset of more complex behaviors and the elimination of previous random and stereotyped behaviors. That is, beyond the time control of molecular advances and cell fate during early gestation, clock genes prepare the fetal organism to exhibit behavioral outputs required for survival after birth. As such, clock genes control the trajectories of emotional clocks and the timely accumulation of emotions aligned to expressed behaviors and their developing complex orientation to human interaction required for emotional survival of the fragile newborn outside the womb.

Thus, the clocks of emotions result in a fetal experience as a sense or feeling by the timing of DNA molecular sensitivity to environmental stimuli at mid-gestation. This capacity is shown in clock gene expression during late gestation, as they begin to oscillate according to their photic reactivity. Previous behaviors which are stereotyped and random represent the embryonic-placental capsule only. The preparation of the fetus to human interaction includes the clocks of emotion onset as elicited by the molecular sensitivity to environmental stimuli and shaping of fetal behaviors in accordance with environmental cycles including the circadian rhythms. Pruning and experienced-shaped molecular events are available by the time molecular sensitivity to the environment starts, rather than being dependent on expressed behavior only. The clocks of emotion are part of the fetal experience which contributes to behavioral feedback loops for pruning purposes. From an evolutionary perspective, the clocks of emotion prepare the fetus to human interaction for survival and growth outside the womb.

References

Ackland, J. F., Ratter, S. J., Bourne, G. L., & Rees, L. H. (1986). Corticotrophin-releasing factor-like immunoreactivity and bioactivity of human fetal and adult hypothalami. *The Journal of Endocrinology, 108*(2), 171–180. https://doi.org/10.1677/joe.0.1080171

Allendoerfer, K. L., & Shatz, C. J. (1994). The subplate, a transient neocortical structure: Its role in the development of connections between thalamus and cortex. *Annual Review of Neuroscience, 17*, 185–218. https://doi.org/10.1146/annurev.ne.17.030194.001153

Arikkath, J. (2012). Molecular mechanisms of dendrite morphogenesis. *Frontiers in Cellular Neuroscience, 6*. https://doi.org/10.3389/fncel.2012.00061

Bauer, P. J. (2006). Constructing a past in infancy: A neuro-developmental account. *Trends in Cognitive Sciences, 10*(4), 175–181. https://doi.org/10.1016/j.tics.2006.02.009

Becker, L. E., Armstrong, D. L., Chan, F., & Wood, M. M. (1984). Dendritic development in human occipital cortical neurons. *Developmental Brain Research, 13*(1), 117–124. https://doi.org/10.1016/0165-3806(84)90083-X

Bresson, J.-L., Clavequin, M.-C., Fellmann, D., & Bugnon, C. (1987). Human corticoliberin hypothalamic neuroglandular system: Comparative immunocytochemical study with anti-rat and anti-ovine corticotropin-releasing factor sera in the early stages of development. *Developmental Brain Research, 32*(2), 241–246. https://doi.org/10.1016/0165-3806(87)90104-0

Carlos, J. D., & O'Leary, D. D. (1992). Growth and targeting of subplate axons and establishment of major cortical pathways [published erratum appears in J Neurosci 1993 Mar;13(3):following table of contents]. *Journal of Neuroscience, 12*(4), 1194–1211. https://doi.org/10.1523/JNEUROSCI.12-04-01194.1992

Cox, K. H., & Takahashi, J. S. (2019). Circadian clock genes and the transcriptional architecture of the clock mechanism. *Journal of Molecular Endocrinology, 63*(4), R93–R102.

Einspieler, C., Prayer, D., & Marschik, P. B. (2021). Fetal movements: The origin of human behaviour. *Developmental Medicine & Child Neurology, 63*(10), 1142–1148. https://doi.org/10.1111/dmcn.14918

Fagard, J., Esseily, R., Jacquey, L., O'Regan, K., & Somogyi, E. (2018). Fetal Origin of Sensorimotor Behavior. *Frontiers in Neurorobotics, 12*. https://doi.org/10.3389/fnbot.2018.00023

Ferber, S. G., Als, H., McAnulty, G., Klinger, G., & Weller, A. (2022). Multi-level hypothalamic neuromodulation of self-regulation and cognition in preterm infants: Towards a control systems model. *Comprehensive Psychoneuroendocrinology, 9*, 100109. https://doi.org/10.1016/j.cpnec.2021.100109

Ferber, S. G., Als, H., McAnulty, G., Peretz, H., & Zisapel, N. (2011). Melatonin and mental capacities in newborn infants. *The Journal of Pediatrics, 159*(1), 99–103.e1. https://doi.org/10.1016/j.jpeds.2010.12.032

Fishell, G., & Rudy, B. (2011). Mechanisms of inhibition within the telencephalon: "Where the Wild Things Are". *Annual Review of Neuroscience, 34*, 535–567. https://doi.org/10.1146/annurev-neuro-061010-113717

Goikolea-Vives, A., & Stolp, H. B. (2021). Connecting the neurobiology of developmental brain injury: Neuronal Arborisation as a regulator of dysfunction and potential therapeutic target. *International Journal of Molecular Sciences, 22*(15), Article 15. https://doi.org/10.3390/ijms22158220

Guadagno, A., Belliveau, C., Mechawar, N., & Walker, C.-D. (2021). Effects of early life stress on the developing basolateral amygdala-prefrontal cortex circuit: The emerging role of local inhibition and Perineuronal nets. *Frontiers in Human Neuroscience, 15*. https://doi.org/10.3389/fnhum.2021.669120

Han, W., & Šestan, N. (2013). Cortical projection neurons: Sprung from the same root. *Neuron, 80*(5), 1103–1105. https://doi.org/10.1016/j.neuron.2013.11.016

Hoshiba, Y., Toda, T., Ebisu, H., Wakimoto, M., Yanagi, S., & Kawasaki, H. (2016). Sox11 balances dendritic morphogenesis with neuronal migration in the developing cerebral cortex. *Journal of Neuroscience, 36*(21), 5775–5784. https://doi.org/10.1523/JNEUROSCI.3250-15.2016

Howland, M. A., Sandman, C. A., & Glynn, L. M. (2017). Developmental origins of the human hypothalamic-pituitary-adrenal axis. *Expert Review of Endocrinology & Metabolism, 12*(5), 321–339. https://doi.org/10.1080/17446651.2017.1356222

Huttenlocher, P. R., & Dabholkar, A. S. (1997). Regional differences in synaptogenesis in human cerebral cortex. *Journal of Comparative Neurology, 387*(2), 167–178. https://doi.org/10.1002/(SICI)1096-9861(19971020)387:2<167::AID-CNE1>3.0.CO;2-Z

Jossin, Y. (2020). Reelin functions, mechanisms of action and signaling pathways during brain development and maturation. *Biomolecules, 10*(6), Article 6. https://doi.org/10.3390/biom10060964

Koutcherov, Y., Mai, J. K., & Paxinos, G. (2003). Hypothalamus of the human fetus. *Journal of Chemical Neuroanatomy, 26*(4), 253–270. https://doi.org/10.1016/j.jchemneu.2003.07.002

Lagercrantz, H., & Changeux, J.-P. (2009). The emergence of human consciousness: from fetal to neonatal life. *Pediatric Research, 65*(3), 255–260. https://doi.org/10.1203/PDR.0b013e3181973b0d

Leader, L. R., Stevens, A. D., & Lumbers, E. R. (2009). Measurement of fetal responses to vibroacoustic stimuli: Habituation in fetal sheep. *Biology of the Neonate, 53*(2), 73–85. https://doi.org/10.1159/000242766

Lee, J. Y., Choi, Y.-H., Cha, J. H., Lee, J. Y., Lee, Y.-J., Kim, B.-H., Kim, I.-K., Lee, J.-M., & Lee, H. J. (2021). Altered asymmetries of the structural networks comprising the fronto-limbic brain circuitry of preterm infants. *Scientific Reports, 11*(1), 1318. https://doi.org/10.1038/s41598-020-79446-0

Lu, D., He, L., Xiang, W., Ai, W.-M., Cao, Y., Wang, X.-S., Pan, A., Luo, X.-G., Li, Z., & Yan, X.-X. (2013). Somal and dendritic development of human CA3 pyramidal neurons from Midgestation to middle childhood: A quantitative Golgi study. *The Anatomical Record, 296*(1), 123–132. https://doi.org/10.1002/ar.22616

Luu, P., & Tucker, D. M. (2023). Continuity and change in neural plasticity through embryonic morphogenesis, fetal activity-dependent synaptogenesis, and infant memory consolidation. *Developmental Psychobiology, 65*(8), e22439. https://doi.org/10.1002/dev.22439

Marchisella, F., Creutzberg, K. C., Begni, V., Sanson, A., Wearick-Silva, L. E., Tractenberg, S. G., Orso, R., Kestering-Ferreira, É., Grassi-Oliveira, R., & Riva, M. A. (2021). Exposure to prenatal stress is associated with an excitatory/inhibitory imbalance in rat prefrontal cortex and amygdala and an increased risk for emotional dysregulation. *Frontiers in Cell and Developmental Biology, 9*, 653384. https://doi.org/10.3389/fcell.2021.653384

Menon, S., & Gupton, S. (2018). Recent advances in branching mechanisms underlying neuronal morphogenesis. *F1000Research, 7*, F1000 Faculty Rev-1779. https://doi.org/10.12688/f1000research.16038.1

Moreno-Juan, V., Filipchuk, A., Antón-Bolaños, N., Mezzera, C., Gezelius, H., Andrés, B., Rodríguez-Malmierca, L., Susín, R., Schaad, O., Iwasato, T., Schüle, R., Rutlin, M., Nelson, S., Ducret, S., Valdeolmillos, M., Rijli, F. M., & López-Bendito, G. (2017). Prenatal thalamic waves regulate cortical area size prior to sensory processing. *Nature Communications, 8*(1), 14172. https://doi.org/10.1038/ncomms14172

Nader, N., Chrousos, G. P., & Kino, T. (2010). Interactions of the circadian CLOCK system and the HPA axis. *Trends in Endocrinology & Metabolism, 21*(5), 277–286. https://doi.org/10.1016/j.tem.2009.12.011

Noda, M., Iwamoto, I., Tabata, H., Yamagata, T., Ito, H., & Nagata, K. (2019). Role of Per3, a circadian clock gene, in embryonic development of mouse cerebral cortex. *Scientific Reports, 9*(1), 5874. https://doi.org/10.1038/s41598-019-42390-9

Pan, Y., & Monje, M. (2020). Activity shapes neural circuit form and function: A historical perspective. *Journal of Neuroscience, 40*(5), 944–954. https://doi.org/10.1523/JNEUROSCI.0740-19.2019

Pepe, G. J., & Albrecht, E. D. (1995). Actions of placental and fetal adrenal steroid hormones in primate pregnancy*. *Endocrine Reviews, 16*(5), 608–648. https://doi.org/10.1210/edrv-16-5-608

Rehman, K. S., Sirianni, R., Parker, C. R., Rainey, W. E., & Carr, B. R. (2007). The regulation of adrenocorticotrophic hormone receptor by Corticotropin-releasing hormone in human fetal adrenal definitive/transitional zone cells. *Reproductive Sciences, 14*(6), 578–587. https://doi.org/10.1177/1933719107307908

Richards, L. J., Koester, S. E., Tuttle, R., & O'Leary, D. D. M. (1997). Directed growth of early cortical axons is influenced by a chemoattractant released from an intermediate target. *Journal of Neuroscience, 17*(7), 2445–2458. https://doi.org/10.1523/JNEUROSCI.17-07-02445.1997

Rosário, M., Schuster, S., Jüttner, R., Parthasarathy, S., Tarabykin, V., & Birchmeier, W. (2012). Neocortical dendritic complexity is controlled during development by NOMA-GAP-dependent inhibition of Cdc42 and activation of cofilin. *Genes & Development, 26*(15), 1743–1757. https://doi.org/10.1101/gad.191593.112

Schoof, E., Girstl, M., Frobenius, W., Kirschbaum, M., Repp, R., Knerr, I., Rascher, W., & Dotsch, J. (2001). Course of placental 11beta-hydroxysteroid dehydrogenase type 2 and 15-hydroxyprostaglandin dehydrogenase mRNA expression during human gestation. *European Journal of Endocrinology, 145*(2), 187–192. https://doi.org/10.1530/eje.0.1450187

Supèr, H., Soriano, E., & Uylings, H. B. M. (1998). The functions of the preplate in development and evolution of the neocortex and hippocampus. *Brain Research Reviews, 27*(1), 40–64. https://doi.org/10.1016/S0165-0173(98)00005-8

Suzuki, N., Zou, Y., Sun, H., Eichel, K., Shao, M., Shih, M., Shen, K., & Chang, C. (2022). Two intrinsic timing mechanisms set start and end times for dendritic arborization of a nociceptive neuron. *Proceedings of the National Academy of Sciences, 119*(45), e2210053119. https://doi.org/10.1073/pnas.2210053119

Tau, G. Z., & Peterson, B. S. (2010). Normal development of brain circuits. *Neuropsychopharmacology, 35*(1), 147–168. https://doi.org/10.1038/npp.2009.115

Taymourtash, A., Schwartz, E., Nenning, K.-H., Sobotka, D., Licandro, R., Glatter, S., Diogo, M. C., Golland, P., Grant, E., Prayer, D., Kasprian, G., & Langs, G. (2023). Fetal development of functional thalamocortical and cortico–cortical connectivity. *Cerebral Cortex, 33*(9), 5613–5624. https://doi.org/10.1093/cercor/bhac446

Tucker, D. M., & Luu, P. (2021). Motive control of unconscious inference: The limbic base of adaptive Bayes. *Neuroscience & Biobehavioral Reviews, 128*, 328–345. https://doi.org/10.1016/j.neubiorev.2021.05.029

Urbanska, M., Blazejczyk, M., & Jaworski, J. (2008). Molecular basis of dendritic arborization [Review]. *Acta Neurobiologiae Experimentalis, 68*(2), Article 2. https://doi.org/10.55782/ane-2008-1695

van Eerdenburg, F. J., & Rakic, P. (1994). Early neurogenesis in the anterior hypothalamus of the rhesus monkey. *Brain Research. Developmental Brain Research, 79*(2), 290–296. https://doi.org/10.1016/0165-3806(94)90134-1

Vasung, L., Huang, H., Jovanov-Milošević, N., Pletikos, M., Mori, S., & Kostović, I. (2010). Development of axonal pathways in the human fetal fronto-limbic brain: Histochemical characterization and diffusion tensor imaging. *Journal of Anatomy, 217*(4), 400–417. https://doi.org/10.1111/j.1469-7580.2010.01260.x

Walkley, S. U., Zervas, M., & Wiseman, S. (2000). Gangliosides as modulators of Dendritogenesis in normal and storage disease-affected pyramidal neurons. *Cerebral Cortex, 10*(10), 1028–1037. https://doi.org/10.1093/cercor/10.10.1028

Wood, J. G., Martin, S., & Price, D. J. (1992). Evidence that the earliest generated cells of the murine cerebral cortex form a transient population in the subplate and marginal zone. *Developmental Brain Research, 66*(1), 137–140. https://doi.org/10.1016/0165-3806(92)90150-U

Yagita, K. (2024). Emergence of the circadian clock oscillation during the developmental process in mammals. *Current Opinion in Genetics & Development, 84*, 102152. https://doi.org/10.1016/j.gde.2024.102152

Yong Ping, E., Laplante, D. P., Elgbeili, G., Hillerer, K. M., Brunet, A., O'Hara, M. W., & King, S. (2015). Prenatal maternal stress predicts stress reactivity at 2½ years of age: The Iowa flood study. *Psychoneuroendocrinology, 56*, 62–78. https://doi.org/10.1016/j.psyneuen.2015.02.015

Part III
INTEGRATION—Bridging Molecular Time-Design and Early Human Experience

Chapter 12
Concluding Chapter: Clock Genes – Beyond Photic Reactivity: The Clock Ticking Behind Very Early Clocks

Abstract The novel techniques of fetal MRI and 3D,4D ultrasound recordings have now made it possible to outline the development of neurites emerging parallelly in preparatory cortices and subcortical regions toward each other, as well as the trajectories of progressing late gestation behavioral organization, originated in earlier random behaviors. Very early clocks work along a human gestational 40-week changing thermodynamic system via timewise alterations in a combinatorial manner, rather than through a simple linear progress. This progress unravels time-specific and tissue-specific functional interplay, where balances of exciting and inhibiting molecular events underlie differential rates of growth. In this chapter, recent findings on the clock genes' control of this combinatorial time-specific and tissue-specific structural development are used to create an integrating model describing their dynamics as "times" of development. This model aims to describe the time-sensitive dynamics of clock genes across developing levels, to show how these dynamics work for morphological advance in a manner that founds the time-dependent changes in embryonic and fetal movements, behavior, and experience. The equation summarizing this model presents a synthesis of thermodynamics, electrochemical principles, and control systems. To show how clock genes determine the time of regulated transition between morphological levels by their expression-suppression ratios, logic gates were employed to show conditions of the regulatory XOR logic gates, created by more primitive Boolean functions. According to this model, to regulate the complex return to balance of their ultradian oscillations, clock genes respond to a given actual value of a structural-related electrochemical condition, following any given structural consolidation in the developing organism. The next desired value of the electrochemical condition, required for the next stage of morphological development, is triggered by the consolidation of the previous stage in a plateau condition. As such, the entire progress of transitions is reflected in a second-order step response. In turn, the electrochemical timely regulated transitions between morphological levels originate the time-sensitive changes in very early behaviors and in the first accumulation of human experiences. It is concluded that clock genes are the clock ticking behind the other very early clocks of human behaviors and experience emergence.

S. Goldstein Ferber, *Very Early Clocks*,
https://doi.org/10.1007/978-3-031-89603-3_12

When is the first human experience identified? What are the early molecular building blocks of this first occurrence? What are the rhythms of their development on the vector of time, and how do they relate to overt, very early behavior? How do the molecular rhythms dictate the timings of morphological development and neural wiring up to elicitation of the first human behavioral repertoire?

The novel techniques of fetal MRI and 3D, 4D ultrasound recordings have now made it possible to outline the development of neurites emerging parallelly in preparatory cortices and subcortical regions toward each other, as well as the trajectories of progressing late gestation behavioral organization originated in earlier random behaviors. Thus, it is not very reasonable any more to conclude that the first human experience is experienced immediately after birth and not before. It is rather more inclusive to conclude that during healthy gestation the clocks of human experience are dependent on intricate but regulated parallel sequences of early molecular events underlying structural growth and neural wiring. In these complex developmental pathways, the timing of events counts more than for each event per se, and even more than the convergence of events solely.

These parallel sequences develop from large levels of entropy, implying simplicity in behavioral outcomes in the embryo through reduced entropy and increased free energy from mid-gestation onwards, with growing complexity in behavioral outcomes late in gestation. Reduced entropy facilitates reoccurrence of a behavioral achievement once it occurred through reversible macro-states in the prenatal thermodynamic system of the human organism (Ferber et al., 2024).

Very early clocks work along a gestational 40-week changing thermodynamic system via timewise alterations in combinations, rather than through a simple linear progress. This progress unravels time-specific functional biological interplay in a combinatorial manner where balances of exciting and inhibiting molecular events underlie differential rates of growth which are time-specific and tissue-specific. Clock genes design the time of available free energy for a behavioral accomplishment by inducing molecular sensitivity to intra- and extrauterine stimulation, beyond their photic reactivity and for shorter and longer timeframes than their known 24-hour cyclicity. As such, clock genes are designed as "timers" to dictate times for other very early clocks.

As described in this book, clock genes dictate the times of molecule allocation and protection of the mitochondria resulting from their very early ultradian oscillations, their modulatory interplay with Ca^{2+} currents, their effects on transcriptomic processes of other genes, and their capacity to activate electron transporters for the conversion of unstable molecules to stable ones.

From the very early stage of mobile Ca^{2+} fluxes in the sperm, clock genes work as "timers," eliciting stable Ca^{2+} currents in the fertilized egg, through polarization of the mitotic and postmitotic events creating the apical domain. The impact of their ultradian oscillations on chromatin, RohA, and F-actin and related enzymatic cascades, which are the early molecular events for apical domain enrichment, and the inner cell population of embryonic development, continues under their control.

Their control continues from genome activation to the birth of the first neuron, the formation of the neural tube, and its closure. The molecular ultradian oscillations of clock genes during embryonic phases result in time-specific electrical fields when they interact with Ca^{2+} currents (Cavieres-Lepe & Ewer, 2021; van der Veen & Gerkema, 2017), thus turning to be an originating source for elicitation, followed by suppression periods, of stereotyped short behavioral cycles, typical in embryos before the gestational timing of the neural tube closure (Einspieler et al., 2021; Fagard et al., 2018).

The clocks of DNA modifications and repair occur as a summation of inverse U-shaped curves of methylations during acute, very early embryonic phases, which peak around mid-gestation, narrowing the range of methylations by the timing of coordination between all tissue-specific transcriptomic clocks, combining upregulation of some genes with downregulation of others. DNA modifications are tissue-specific and time-specific, and their shifting time parallels the time of emergence of very early human capacities at mid-gestation, upon reaching the borderline of viability. Their optimal range narrows from mid-gestation to infancy (Ferber et al., 2022). This complex process develops under the governance of clock genes that were found to exert an overarching architecture (Takahashi, 2017) affecting the entire genome's transcription in a time-sensitive manner.

The birth of the first neurons and the formation of the neural tube are not a sufficient development to elicit organized and regulated human behavior, as this depends on coordination of neural signaling achieved later in gestation. Neural signaling, in turn, develops towards a coordinated status as a result of myelination. Myelination of axons is the last stage of white matter development which overlaps with early stages of pruning not elicited by experience. Myelination starts early during embryonic stages and continues to early infancy. Myelin-producing cells respond to intra- and extracellular time cues for the timing of each of their waves and for progressing through their stages of maturation. Timely myelination contributes to accuracy of neural firing more than to velocity per se. This accuracy results from myelinating cells' reaction to clock genes' transcription and their impact on the redistribution of ion channels, especially those of sodium and calcium fluxes (Freeman et al., 2016).

Beyond the regulation of ion distributions and transcriptions of other genes, clock genes work to mark timings for the entire organism through the guardians of the balance between protecting and endangering free radicals, that requires on-off switches on the vector of time. This enables clock genes dynamics to protect the mitochondria's changing needs for energy production along gestation. The rhythmicity of clock gene expression and their elicitation and control of molecular processes, involving electron transport and transformation of unstable molecules to stable ones (Hansen et al., 2020), secures the healthy progress of gestation through differential levels of oxygenation at each given time. The relation between velocity and timing of clock gene oscillations represents the match-mismatch between the resulting rhythmicity of free radicals and the rate of organogenesis at any given time along gestation. That is, clock genes, through their on-off switches of oxygenation during gestation, control the balance between oxidative stress and antioxidative defenses.

It is accepted that generating the human six-layered cerebral cortex, which serves for the elicitation of more regulated behaviors, is dependent on integrating signals from various cell types. This accomplishment is dependent on the ionic gradients of the cell membrane potentials, and accurate function of the ATPases. That is, the molecular rhythms that lead firing rates and their timing during embryonic life develop through mid-gestation to afford the elicitation of early fetal self-regulatory behaviors and their return to balance after mounting arousal. ATPase events are reversible and shaped by behavior from mid-gestation onwards. At the time window of mid-gestation, ATPases reduce the dominance of excitation and start to balance it under the control of clock genes (Smith et al., 2021), which by then gradually shift from ultradian to circadian oscillations. ATPases function as the early flexible basis of the emerging self-regulated behaviors, while timing of changes in ATPase expression is controlled by clock genes (Qi et al., 2023).

As described in this book in length, in recent years, changes from early embryonic stereotyped behaviors to stimulus-related fetal behaviors have been recorded following mid-gestation. These include capacities such as habituation and orienting response, which incorporate a transitory quality of awareness. Following embryonic rapid phases of development, a rise in inhibitory factors is apparent, contributing to the onset of excitation-inhibition balance. Accumulation of more organized, intentional, and stimulus-related fetal behaviors occurs with this rise of inhibitory factors. This rise of inhibitory factors at mid-gestation is controlled by Cl^- mobilization of Ca^{2+} cell entries towards more balanced frequencies of the neural cell action potentials, as part of the cell's increasing readiness to partner in postsynaptic transmission and wiring. The same timing of changes in the Cl^- dynamics and in the type of clock gene oscillations (shifting from ultradian oscillations only to circadian rhythms) coincides with the onset of experience-shaped synaptic changes at mid-gestation. From this timing through birth, feedback loops exist between the behavioral and the molecular levels, resulting in the onset of coordinated firing of neural networks and corresponding stimulus-responsive behavioral outcomes.

Clock genes govern the rhythms of inhibitory factors though their heterodimerizing effects and modulation of the excitatory Ca^{2+} entries to the cell. The capacity of clock genes to converge unstable molecules to stable ones reduces the amounts of oxidative stress and elicits antioxidative defenses, thus promoting the rise of excitation-inhibition balance during the transition of mid-gestation. By their noted capacities, clock genes protect the mitochondrial functions to be gestational age-appropriate for the metabolic demands; thus, their functioning rhythms protect the advance in structural growth. This in turn supports more efficient fetal recovery from unpleasant intrauterine conditions and inhibition of behaviors aligned to these conditions.

Thus, at mid-gestation, when the onset of excitation-inhibition rhythms meets structural accomplishments appropriate for this gestational age, through a complex of inhibitory factors, elicitation of intentional behaviors occurs on the expense of earlier more frequent random behaviors. Behavioral regulation thus corresponds to the clock genes' rhythms of arousal buildup and return to excitation-inhibition balance.

Differential time-sensitive and tissue-specific combinatorial rates of growth within brain regions and subregions are shown in recent findings (Arlotta et al., 2005). This structural intricate development is based on early gestational functions of ionic regulation, especially chloride, sodium, and potassium, which in turn is governed by ultradian oscillations of clock genes for timely ion transport and cell osmotic regulation, as described in this book. The time window of mid-gestation, associated with the borderline of viability and the onset of pleasure-like fetal behaviors, is suggested as the time of convergence between earlier molecular building blocks and structural-functional network establishment.

Cortical and subcortical regions develop early structural volumes toward each other, and their mutual embrace elicits the components of the alert state and more complex behavioral phenomena later in gestation. These processes develop while residual physiological and cortical excitability, which follows the behavioral execution of a function, decreases from mid-gestation onward, showing more coordination of network wiring and feedback loops from the behavioral to the molecular level. At the time window of mid-gestation, any remaining differentiation starts to overlap with pruning processes. As shown in the data included in this book, from mid-gestation onward, this molecular basis, stratified by clock genes, interacts with the environment too, in a manner that any fetal acquisition of an executive function cannot be predicted by either the clocks or the environmental stimuli only.

This complex of structural progress results from the combinatorial time-specific and tissue-specific manner of neural network development into the cortical ability to reach short states of awakening. Once these brief states have occurred, the alert state span may widen in later repetitions of this condition often described as a paradoxical alpha wave, with the return of the alpha wave to the EEG pattern. Later in gestation, the EEG alpha wave, representing awake and partial awareness states, occurs on the expense of delta waves detected during earlier stages of gestation.

Thus, as shown here, gestation is not a simple linear process. It is about controls and balances during healthy gestation, while clock genes design the timing of DNA sensitivity to changes across atomic, molecular, cellular, networks, and structural advances in regional and inter-regional levels. From mid-gestation onward, clock genes design the molecular sensitivity to environmental stimuli. Pruning and experienced-shaped molecular events are available by the time molecular sensitivity to the environment starts.

12.1 The Model

How do the molecular rhythms dictate the timings of morphological development and neural wiring up to elicitation of the first human behavioral repertoire? This model aims to describe the time-sensitive dynamics of clock genes across developing levels in the human embryo and fetus and to show how these dynamics work for morphological advance and its neural connectivity, founding the emergence of human behavior. As such, this model aims to show that the oscillatory dynamics of clock genes trigger numerous parallel sequences of morphological events in a time-sensitive manner, which in turn converge into one behavioral achievement at a time.

It is assumed that clock genes function as a thermodynamic system. Viewing molecular events underlying overt time-dependent and condition-dependent behaviors as a thermodynamic system has been suggested earlier (Ferber et al., 2023). This view implies a free energy-entropy matrix, where free energy reflects the work (useful energy) which a system can perform at a constant temperature, that is, the energy quantity for a transition or change within and between the developing levels of the embryo and fetus. It is also assumed that chemical transition or change in the dynamics of clock genes involves electrical foundations, and any progress appears as a step response from zero to 1, followed by a time period of plateau in oscillatory functions. The following equation integrates these principles. This synthesis bridges thermodynamics, dynamic control systems theory, and electrochemical principles, making it ideal for modeling electrochemical organisms or systems which are alerted by transient and complex time-responsive oscillations such as the molecular advancing events triggered by clock genes along gestation.

1. **Thermodynamic laws**
   - The Gibbs free energy equation is the starting point.
   - Time dependency and enthalpy-entropy interplay define the thermodynamic baseline.
2. **Control systems theory**
   - The second-order step response function describes time-dependent behavior (oscillatory or non-oscillatory).
3. **Electrochemical principles**
   - Faraday's constant F introduces the effect of electrical work on free energy.
4. **Heaviside function**
   - Used to model step inputs, enabling external changes (e.g., sudden voltage application).
5. **Superposition principle**
   - Thermodynamic, oscillatory, and electrochemical components are combined, each contributing to the system's time-dependent response.

12.2 Explanation: How this Synthesis Was Created

The final equation is:

$$\Delta G(t) = \left[(\Delta H_0 - T_0 \Delta S_0) + u_0 H(t)\right]\left(1 - e^{-\frac{\zeta\omega_n t}{\sqrt{1-\zeta^2}}} \sin(\omega_d t + \varnothing)\right) - nFE(t)$$

This equation is a synthesis of thermodynamics, electrochemical principles, and dynamic system step response concepts. Here is a staged breakdown of how the different components and known equations were combined to create this model:

1. **Start with the Gibbs Free Energy Equation**

   Assuming that the gestational timely development is dependent on the thermodynamic free energy entropy matrix, including time indication for cell division, proliferation, and migration (Ferber et al., 2023), the basic form of the Gibbs free energy equation is:

$$\Delta G = \Delta H - T\Delta S \tag{12.1}$$

where:

- ΔG: Gibbs free energy (determines system spontaneity).
- ΔH: Enthalpy (total energy of the system).
- $T\Delta S$: Entropic contribution (energy lost to disorder).

   This is the thermodynamic foundation of the equation.

2. **Add Time Dependency to Gibbs Free Energy Equation**

   Gestational development is not a simple linear process but rather occurs as a time-dependent and tissue-dependent increasing complexity. To account for this:

- Replace T and S with their initial values ($T = 0$, $\Delta S = 0$ to define a baseline).
- Add a **step function** to model external perturbations:
  - u0H(t), where H(t) is the **Heaviside step function**, which introduces a sudden change at $t = 0$ (e.g., a voltage step, change in temperature, etc.).

   Now the equation becomes:

$$\Delta G(t) = \left[(\Delta H_0 - T_0 \Delta S_0) + u_0 H(t)\right] f(t) \tag{12.2}$$

where f(t) is a time-dependent response function.

3. **Incorporate Second-Order System Dynamics**

   The time-dependent behavior of a physical and electrochemical system such as gestational development involves increasing complexity as gestation advances. It occurs as a timely sudden change followed by a time period of system consolidation of the change, represented by a plateau. This type of staged progress was shown in sudden changes in embryonic and later fetal behaviors detected by 3D and 4D ultrasound recordings (new behaviors follow elimination of earlier behaviors) (Einspieler et al., 2021; Fagard et al., 2018), which were shown in this book to be the output of time-dependent molecular advance and rearrangements of thermodynamic properties. Therefore, thermodynamic responses can be termed as a transient response of a second-order system (Dorf, 1980). The second-order differential equations describe:

- Damped oscillations (under damped systems).
- Critical damping (fastest return to equilibrium without oscillations).
- Overdamped systems (slow return without oscillations).

Damped oscillations are considered to describe the timely change from overshooting of a molecular step response towards a consolidation stage described by a plateau which follows.

Thus, the step response function for a second-order system is:

$$f(t) = 1 - e^{-\frac{\zeta\omega_n t}{\sqrt{1-\zeta^2}}} \sin(\omega_d t + \varnothing) \tag{12.3}$$

where:

e$^{\cdot}$**:** This represents an **exponential decay function**, which controls how quickly the oscillations decrease over time. The exponent determines the rate of decay.

ζ: The **damping ratio**, a dimensionless parameter that determines how oscillations decay over time. That is, the **damping ratio**, which influences the decay rate. A higher ζ leads to faster decay. It appears in the numerator, meaning it directly affects the speed of decay.

- If $\zeta < 1$, the system is **underdamped**, and oscillations occur.
- If $\zeta = 1$, the system is **critically damped**, with no oscillation but the fastest return to equilibrium.
- If $\zeta > 1$, the system is **overdamped**, and it returns to equilibrium slowly without oscillation.

- ωn: Natural frequency. The **undamped natural frequency**, which is the frequency of oscillations if no damping were present. That is, the **natural frequency**, which determines how fast the system would oscillate without damping. It scales the time variable.

- **ωd:** The **damped natural frequency**, which represents the oscillations frequency of the system after accounting for energy dissipation (damping).

ωd = ωn1 − ζ2 : Damped frequency.

- This term adjusts the decay rate by accounting for underdamping effects.
- When $\zeta < 1$, the denominator reduces the overall effect of ζωnt leading to a slower decay.

- When $\zeta \rightarrow 1$, the denominator approaches zero, meaning a critically damped or overdamped response.
- t: Represents time; That is, the system evolves as time increases.
- ϕ: Phase shift due to initial conditions.

This function describes the system's approach to equilibrium following a sudden perturbation.

4. **Incorporate Electrochemical Effects (Faraday's Constant)**

The timely oscillations of clock genes affecting primarily electron transporters and cotransporters to covert unusable molecules to stable ones present the underlying molecular progress of a gestational time-dependent development as an electrochemical system, which implies that the Gibbs free energy is affected by electrochemical work**.** This is represented here by the Faraday constant, showing the amount of charge for a chemical substance to be electrolyzed. Electrolocation is assumed for any step response development along gestation. This is described as:

$$\Delta G = -nFE \tag{12.4}$$

where:

- n: Number of moles of electrons transferred.
- F: Faraday's constant (96,485 Coulomb/mol).
- E: Electrochemical potential (voltage).

Adding this to the time-dependent system gives:

$$\Delta G(t) = \left[\left(\Delta H_0 - T_0 \Delta S_0\right) + u_0 H(t)\right] f(t) - nFE(t) \tag{12.5}$$

5. **Final Equation**

Substituting the second-order step response f(t) into the equation gives:

$$\Delta G(t) = \left[\left(\Delta H_0 - T_0 \Delta S_0\right) + u_0 H(t)\right] \left(1 - e^{-\frac{\zeta \omega_n t}{\sqrt{1-\zeta^2}}} \sin\left(\omega_d t + \varnothing\right)\right) - nFE(t) \tag{12.6}$$

12.3 Why This Equation Works

- **Thermodynamic foundation:** Captures initial energy states.
- **Dynamic response:** Accounts for how systems evolve towards equilibrium in a time-sensitive manner after their original status is interfered.
- **Electrochemical effects:** Includes electrical work done by or on the system, making it relevant to electrochemical processes.

This equation arises questions related to the transitions of chemoelectrical statuses between different morphological levels ranging from the atomic level to molecular, cellular, network, regional, and inter-regional levels and how the dynamics of clock genes afford these timely morphological advances along gestation. To determine the time of transition and to regulate their complex return to balance of their oscillatory reactions, clock genes respond to a given structural-related electrochemical condition following its consolidation in the developing organism. How do these dynamics work? It is given that the fertilized egg develops from the atomic originating level (zero- to eight-cell stage) to molecular (two- to 16-cell stage), cellular [from16-cell stage (gestational day 4) up to 16 weeks of gestation], network wiring (3–23 weeks, gestational age), distinct regional growth (12–24 weeks, gestational age), and inter-regional connectivity (19–40+ gestational age, continuing postnatally). The different levels are added along gestation in parallel sequences and partial overlap on the vector of time (Fig. 12.1).

Adapted from control systems theory, to design the time of any transition in gestation, clock genes compare between an actual value of structural-related electrochemical status and a desired value which includes the electrochemical status that is required for the next structural advance. Their expression-suppression ratio responds to the free energy-entropy matrix at a given time for a given structural condition. This creates numerous conditions which are described here as logic gates.

XOR gates have been shown to function as a regulated transition between levels (Ferber, 2008; Ferber et al., 2023). From the primitive logic gates, AND gates (for expression) and NOT gates (for suppression), in a regulated interplay by the primitive Boolean function of OR, XOR gates are created. The interplay of AND and NOT primitive gates creates a regulated transition by XOR gate between levels (Fig. 12.2). Figure 12.2 describes how the interplay between AND logic gates (showing expression conditions of clock genes) and NOT logic gates (showing suppression conditions of clock genes) is regulated by OR logic gates and resumes in XOR logic gates which afford a regulated electrochemical transition between levels (Fig. 12.2). As shown in Table 12.1, the transient oscillations are described as differential combinations of logic gates minimized into XOR gate output conditions, shown in minimized Table 12.2, reflecting timely conditions of XOR gate and implying decreased oscillatory range.

To summarize the model, each developmental level is added to the previously achieved structural-related electrochemical condition in partially overlapping time windows. The levels accumulate as the time of gestation evolves. The availability of feedback loops between structural-related electrochemical conditions and the oscillatory complex of clock genes along gestation advances until the time of approaching birth. The feedback from the electrochemical condition for any given structural status of the developing organism determines the amount of free energy required for the next structural stepwise development. In this manner, the amount of the desired free energy for the next developmental step is compared by clock genes to the actual availability of free energy, through their sensitivity to the molecular events along gestation and their direct interactions with metabolic processes and protein

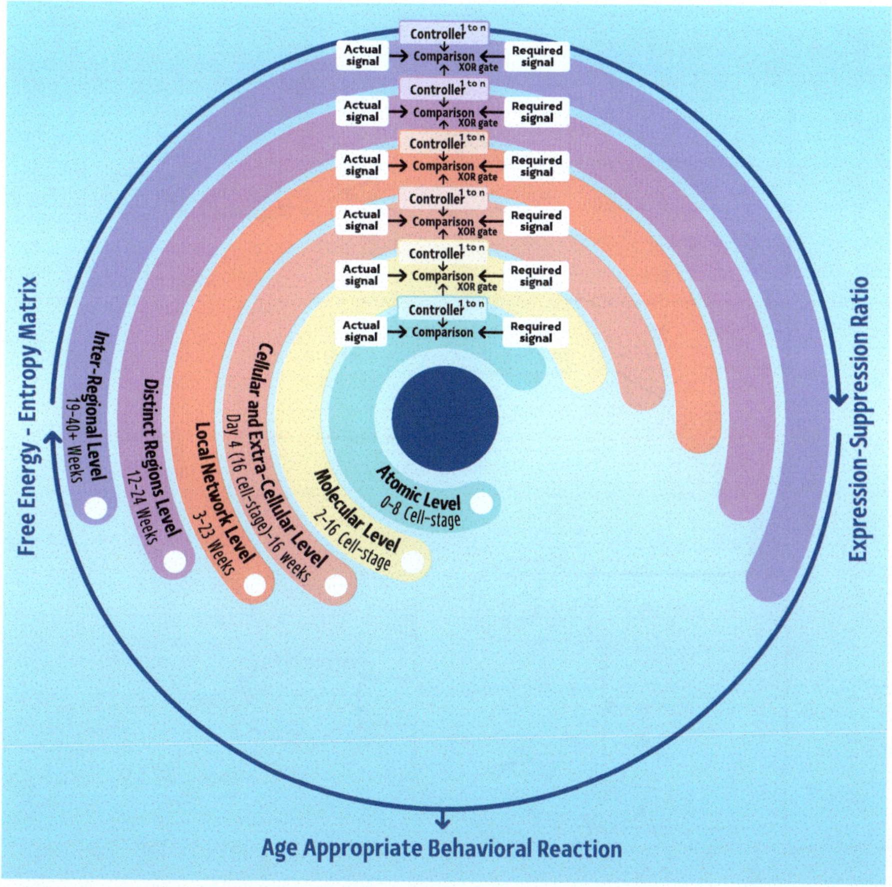

Fig. 12.1 A diagram of the control systems model of clock gene healthy function on the vector of time along gestation is shown in parallel sequences using XOR gate outputs between time-dependent developing levels (controllers). These parallel sequences are shown as resulting from clock gene expression-suppression ratio responding to structural-related electrochemical conditions at each level, at a given gestational age.

production, via their work with chromatin. The cue to oscillate and jump to the next developmental stage is available to clock genes by the comparison of the actual value to the desired value. This comparison is resolved in the feedback loops between a structural-related electrochemical condition and coded electrochemical requirements for the next level or developmental stage. The next stage of development is triggered by consolidation of the previous stage in a plateau condition. The entire progress of transitions through the creation of XOR gates between morphological levels is reflected in a second-order step response function (Goldstein Ferber et al., 2022). Thus, in healthy gestation clock genes elicit regulated transitions and develop in increasing complexity of oscillatory second-order transient responses

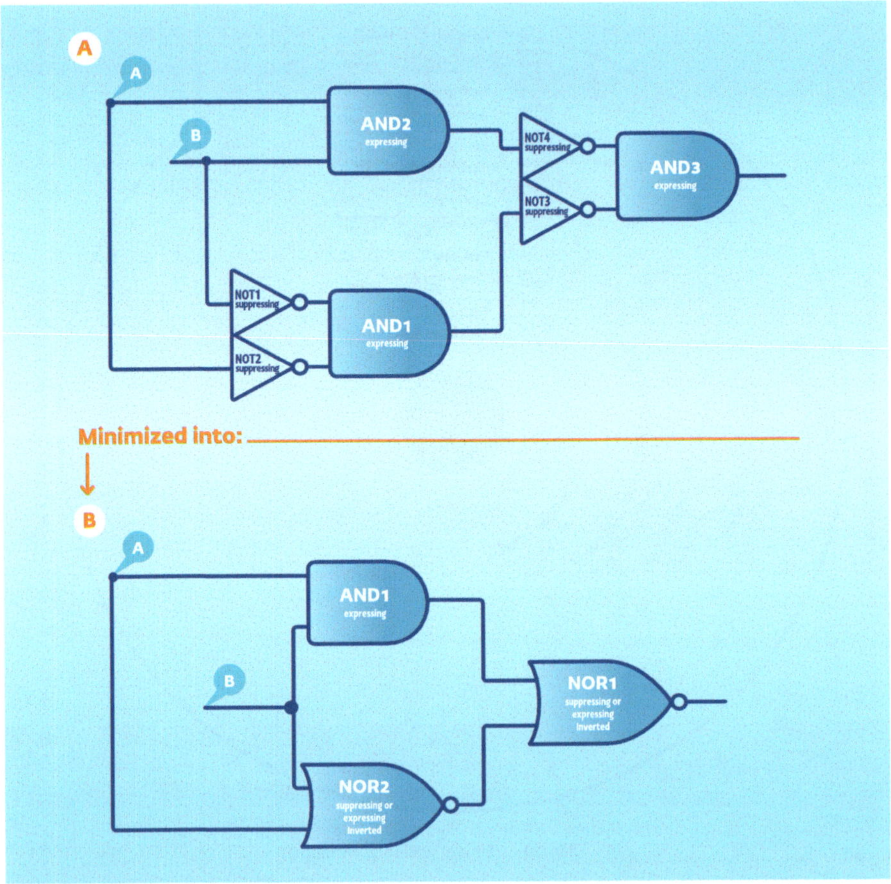

Fig. 12.2 The logic interplay for the creation of XOR gate Boolean function by primitive building blocks of AND gates (which represent clock gene expression) and NOT gates (which represent clock gene suppression) regulated by the OR gates. (**a**) The entire logic version of the interplay is shown in the upper part of the figure. (**b**) In the lower part, the output of the NOR1 gate is a XOR = 1 gate function.

processing numerous primitive logic gates into XOR gate transitions followed by plateau phases, emphasizing that healthy gestation requires regulated transitions.

It is emphasized that the developmental jumps between morphological levels occur in parallel sequences (Ferber et al., 2023) while each sequence starts at a different point of time. It is also noted that structural advances are possible by the transcriptomic impacts of clock genes on the entire genome (Takahashi, 2017) and that clock genes retain the crucial role of acting as "timers" of the organismal development. Any proceeding step is indeed coded in the genetic code, but its actualization at the appropriate time in gestation depends on these structural-related

Table 12.1 Truth table for five transitions through six levels (represents logic conditions of any detected structural-related electrochemical condition). Reference list for XOR gate output=1 in 5 parallel sequences between 6 levels (Table 12.1 and 12.2): 1=sequence between A and B; 2=sequence between B and C; 3= sequence between C and D; 4=sequence between D and E; 5=sequence between E and F.

XOR output (see reference list above)	Atom level: A	Molecular level: B	Cellular level: C	Network level: D	Regional level: E	Inter-regional level: F
	0	0	0	0	0	0
5	0	0	0	0	0	1
4,5	0	0	0	0	1	0
4	0	0	0	0	1	1
3,4	0	0	0	1	0	0
3,4,5	0	0	0	1	0	1
3,5	0	0	0	1	1	0
3	0	0	0	1	1	1
2,3	0	0	1	0	0	0
2,3,5	0	0	1	0	0	1
2,3,4,5	0	0	1	0	1	0
2,3,4	0	0	1	0	1	1
2,4	0	0	1	1	0	0
2,4,5	0	0	1	1	0	1
2,5	0	0	1	1	1	0
2	0	0	1	1	1	1
1,2	0	1	0	0	0	0
1,2,5	0	1	0	0	0	1
1,2,4,5	0	1	0	0	1	0
1,2,4	0	1	0	0	1	1
1,2,3,4	0	1	0	1	0	0
1,2,3,4,5(*)	0	1	0	1	0	1
1,2,3,5	0	1	0	1	1	0
1,2	0	1	0	1	1	1
1,3	0	1	1	0	0	0
1,3,5	0	1	1	0	0	1
1,3,4,5	0	1	1	0	1	0
1,3,4	0	1	1	0	1	1
1,4	0	1	1	1	0	0
1,4,5	0	1	1	1	0	1
1,5	0	1	1	1	1	0
1	0	1	1	1	1	1
1	1	0	0	0	0	0
1,5	1	0	0	0	0	1
1,4,5	1	0	0	0	1	0

(continued)

Table 12.1 (continued)

XOR output (see reference list above)	Atom level: A	Molecular level: B	Cellular level: C	Network level: D	Regional level: E	Inter-regional level: F
1,4	1	0	0	0	1	1
1,3,4	1	0	0	1	0	0
1,3,4,5	1	0	0	1	0	1
1,3,5	1	0	0	1	1	0
1,3	1	0	0	1	1	1
1,2,3	1	0	1	0	0	0
1,2,3,5	1	0	1	0	0	1
1,2,3,4,5(*)	1	0	1	0	1	0
1,2,3,4	1	0	1	0	1	1
1,2,4	1	0	1	1	0	0
1,2,4,5	1	0	1	1	0	1
1,2,5	1	0	1	1	1	0
1,2	1	0	1	1	1	1
2	1	1	0	0	0	0
2,5	1	1	0	0	0	1
2,4,5	1	1	0	0	1	0
2,4	1	1	0	0	1	1
2,3,4	1	1	0	1	0	0
2,3,4,5	1	1	0	1	0	1
2,3,5	1	1	0	1	1	0
2,3	1	1	0	1	1	1
3	1	1	1	0	0	0
3,5	1	1	1	0	0	1
3,4,5	1	1	1	0	1	0
3,4	1	1	1	0	1	1
4	1	1	1	1	0	0
4,5	1	1	1	1	0	1
5	1	1	1	1	1	0
	1	1	1	1	1	1

Table 12.2 Minimization table (*represents regulated transitions between structural-related electrochemical conditions shown in XOR gate output)

XOR output (see reference list above)	Atom level: A	Molecular level: B	Cellular level: C	Network level: D	Regional level: E	Inter-regional level: F
1,2,3,4,5*	0	1	0	1	0	1
1,2,3,4,5*	1	0	1	0	1	0

electrochemical conditions calculated by clock genes and their responsive oscillations to elicit morphological advances on time. The electrochemical timely regulated transitions between morphological levels create the foundation of the time-sensitive changes in very early behaviors and the first human experiences as they evolve on the basis of these behavioral changes.

12.4 Conclusions

It is concluded that clock genes are sensitive to structural-related electrochemical conditions of the developing organism, while the timing, range, and regulatory complexity of their expression-suppression oscillatory functions are determined by the conditions they detect. As such, the mystery of very early clocks lies beyond photic reactivity of clock genes. It is further concluded that clock genes are the clock ticking behind the other very early clocks of human behavioral and emotional emergence. From an evolutionary perspective, life on Earth and its cyclicity shaped how and when very early clocks function to prepare the human organism for behavioral and emotional survival outside the womb.

References

Arlotta, P., Molyneaux, B. J., Chen, J., Inoue, J., Kominami, R., & Macklis, J. D. (2005). Neuronal subtype-specific genes that control corticospinal motor neuron development in vivo. *Neuron, 45*(2), 207–221. https://doi.org/10.1016/j.neuron.2004.12.036

Cavieres-Lepe, J., & Ewer, J. (2021). Reciprocal relationship between calcium signaling and circadian clocks: Implications for calcium homeostasis, clock function, and therapeutics. *Frontiers in Molecular Neuroscience, 14*. https://doi.org/10.3389/fnmol.2021.666673

Dorf, R. C. (1980). *Modern control systems*. Addison-Wesley.

Einspieler, C., Prayer, D., & Marschik, P. B. (2021). Fetal movements: The origin of human behaviour. *Developmental Medicine & Child Neurology, 63*(10), 1142–1148. https://doi.org/10.1111/dmcn.14918

Fagard, J., Esseily, R., Jacquey, L., O'Regan, K., & Somogyi, E. (2018). Fetal origin of sensorimotor behavior. *Frontiers in Neurorobotics, 12*. https://doi.org/10.3389/fnbot.2018.00023

Ferber, S. G. (2008). The concept of coregulation between neurobehavioral subsystems: The logic interplay between excitatory and inhibitory ends. *Behavioral and Brain Sciences, 31*(3), 337–338. https://doi.org/10.1017/S0140525X08004123

Ferber, S. G., Weller, A., & Soreq, H. (2023). Control systems theory revisited: New insights on the brain clocks of time-to-action. *Frontiers in Neuroscience, 17*. https://doi.org/10.3389/fnins.2023.1171765

Ferber, S. G., Weller, A., & Soreq, H. (2024). Boltzmann's theorem revisited: Inaccurate time-to-action clocks in affective disorders. *Current Neuropharmacology, 22*(11), 1762–1777. https://doi.org/10.2174/1570159X22666240315100326

Ferber, S. G., Als, H., McAnulty, G., Klinger, G., & Weller, A. (2022). Multi-level hypothalamic neuromodulation of self-regulation and cognition in preterm infants: Towards a control systems model. *Comprehensive Psychoneuroendocrinology, 9*, 100109.

Freeman, S. A., Desmazières, A., Fricker, D., Lubetzki, C., & Sol-Foulon, N. (2016). Mechanisms of sodium channel clustering and its influence on axonal impulse conduction. *Cellular and Molecular Life Sciences, 73*(4), 723–735. https://doi.org/10.1007/s00018-015-2081-1

Goldstein Ferber, S., Weller, A., Ben-Shachar, M., Klinger, G., & Geva, R. (2022). Development of the ontogenetic self-regulation clock. *International Journal of Molecular Sciences, 23*(2), Article 2. https://doi.org/10.3390/ijms23020993

Hansen, J. M., Jones, D. P., & Harris, C. (2020). The redox theory of development. *Antioxidants & Redox Signaling, 32*(10), 715–740. https://doi.org/10.1089/ars.2019.7976

Qi, J., Mo, F., An, N. A., Mi, T., Wang, J., Qi, J.-T., Li, X., Zhang, B., Xia, L., Lu, Y., Sun, G., Wang, X., Li, C.-Y., & Hu, B. (2023). A human-specific De novo gene promotes cortical expansion and folding. *Advanced Science, 10*(7), 2204140. https://doi.org/10.1002/advs.202204140

Smith, R. S., Florio, M., Akula, S. K., Neil, J. E., Wang, Y., Hill, R. S., Goldman, M., Mullally, C. D., Reed, N., Bello-Espinosa, L., Flores-Sarnat, L., Monteiro, F. P., Erasmo, C. B., Pinto e Vairo, F., Morava, E., Barkovich, A. J., Gonzalez-Heydrich, J., Brownstein, C. A., McCarroll, S. A., & Walsh, C. A. (2021). Early role for a Na+,K+-ATPase (ATP1A3) in brain development. *Proceedings of the National Academy of Sciences, 118*(25), e2023333118. https://doi.org/10.1073/pnas.2023333118

Takahashi, J. S. (2017). Transcriptional architecture of the mammalian circadian clock. *Nature Reviews Genetics, 18*(3), 164–179. https://doi.org/10.1038/nrg.2016.150

van der Veen, D. R., & Gerkema, M. P. (2017). Unmasking ultradian rhythms in gene expression. *The FASEB Journal, 31*(2), 743–750. https://doi.org/10.1096/fj.201600872R

The manufacturer's authorised representative in the EU is Springer Nature Customer Service Centre GmbH, Europaplatz 3, 69115 Heidelberg, Germany. If you have any concerns regarding our products, please contact ProductSafety@springernature.com

Printed and bound by CPI Group (UK) Ltd, Croydon, CR0 4YY
05/10/2026
02228159-0001